Advanced Information and Knowledge Processing

Series Editors

Lakhmi C. Jain
Adelaide, South Australia, Australia

Xindong Wu
University of Vermont Dept. Computer Science, Burlington, Vermont, USA

Information systems and intelligent knowledge processing are playing an increasing role in business, science and technology. Recently, advanced information systems have evolved to facilitate the co-evolution of human and information networks within communities. These advanced information systems use various paradigms including artificial intelligence, knowledge management, and neural science as well as conventional information processing paradigms. The aim of this series is to publish books on new designs and applications of advanced information and knowledge processing paradigms in areas including but not limited to aviation, business, security, education, engineering, health, management, and science. Books in the series should have a strong focus on information processing - preferably combined with, or extended by, new results from adjacent sciences. Proposals for research monographs, reference books, coherently integrated multi-author edited books, and handbooks will be considered for the series and each proposal will be reviewed by the Series Editors, with additional reviews from the editorial board and independent reviewers where appropriate. Titles published within the Advanced Information and Knowledge Processing series are included in Thomson Reuters' Book Citation Index.

More information about this series at http://www.springer.com/series/4738

Tshilidzi Marwala

Artificial Intelligence Techniques for Rational Decision Making

 Springer

Tshilidzi Marwala
Faculty of Engineering and the Built
   Enviroment
University of Johannesburg
Auckland Park
South Africa

ISSN 1610-3947
ISBN 978-3-319-38298-2   ISBN 978-3-319-11424-8 (eBook)
DOI 10.1007/978-3-319-11424-8
Springer Cham Heidelberg New York Dordrecht London

Printed on acid-free paper

Springer is part of Springer Science+Business Media (www.springer.com)

Foreword

Philosophy and Mathematics sections in libraries worldwide, are not in short supply of titles on rationality. When rationality delves into decision making, the offers easily fold and this is relatively easy to figure out because these topics are not only theoretically intriguing but encompass endless practical implications. For that reason, discussions on rationality and decision making will not to be exhausted any time soon. Nonetheless, the ideas and skilful approach taken by Tshilidzi Marwala, in his new book, "Artificial Intelligence for Rational Decision Making", to say the least, are highly refreshing and inspiring. The text is indeed quite readable, very stimulating and often teasing. The contextual elements, down-to-earth examples, and extensive list of further references are a rare asset to the readers who certainly will enjoy the broad overview on Rational Decision Making provided in Chap. 1.

Rationality can be tackled and discussed from many different perspectives. In a rather simplistic manner, it can be seen as the "distance" from beliefs to reasons within a coherent system. Regardless of the reader's purpose for reading this book, be it for further understanding of the topic and/or for application purposes, the difficult formalization of such a subject and the lucid explanations provided here for all circumventing concepts are not only necessary but highly welcome. For example, in Chaps. 2 and 3, rough set and support vector machines, respectively, are instrumental for creating causal and correlation functions. Analogously, missing data and counterfactuals, of Chaps. 4 and 5 are implemented using sound artificial intelligence methods and this is precisely where this book excels as it offers formal but creative and palatable perspectives on rationality.

The key-concepts of "flexibly-bounded rationality", "flow of information" and "marginalization of irrelevance", presented in Chaps. 6 and 7, are seminal and master-strokes for one more securely now (together with the other previous propositions) to be able to ascribe either causality or mere correlation i.e., some information has to flow (or emanate-flexibly) among relating concepts in order for any cause/correlation to be reputed. A generalization, so to say on rational decision making is then on offer in Chap. 8–group decision, when ensembles and mixtures of techniques are concocted graciously.

In all, it was a most fortunate and enjoyable task for me to write this foreword as the book surely fills a gap in the long-standing and profuse discussions about rationality. I personally appreciated greatly how the author engaged on the key aspects, without taking camps, but rather, offering new avenues for one to take up the subject, yet using sound methods from artificial intelligence. Remarkable is also the direction put forward for feeding the reader's thoughts, regardless of whether they are for abstract or real purposes. Concomitant ideas comprised in this book quite certainly will inspire scientists and engineers to produce computable embodiments of rationality. Interested readers will surely profit, seamlessly, to readily delve into the associated logic-philosophical debate, mathematically formulated and computationally exemplified.

I conclude by assuring the lucky readers that the topic of rationality for decision making will be reasoned very differently in their minds after learning the complexly-simple rationale here conceived by Tshilidzi Marwala. I can foresee that after exposure to this text: philosophers will feel the need to (re-)think, scientists will be urged to (re-)write, engineers will wish to (re-)build and for sure lecturers will be highly rewarded when they (re-)ad.

Recife   Fernando Buarque de Lima Neto
October 2014

Preface

This book is on decision making, which is a complex process, that has confounded intellectuals since time immemorial. For instance, to make decisions rationally, one must take into account that others may be acting irrationally. In this instance, the rational decision that is taken will feature into account the irrational reaction from the other role player. As an example, a doctor prescribes medicine to a superstitious individual. That decision making process (rational) of prescribing medicine might take into account of the individual's superstitious inclinations (irrational) and, therefore, also prescribes that the patient should be supervised on taking the medication. So in essence the decision taken has factored into account the possibility the patient might act irrationally.

This book proposes that the basic elements of rational decision making are causality and correlation functions. Correlation is important because it fills in the gaps that exist in information needed for decision making because of the limitation of information available in decision making. Causality is important in decision making because it is through causal loops (if this then act so) that an appropriate cause of action is chosen from many possible causes of actions, based on the principle of the maximization of utility.

This book defines a causal function and builds it using rough sets and successfully applies it for decision making. This book defines a correlation function and builds it using support vector machines and applies these to model epileptic activity. Missing data approach which is based on the multi-layer perceptron autoassociative network and genetic algorithm is proposed and applied for decision making in antenatal data set and is then used for HIV prediction. Furthermore, this book introduces the theory of rational counterfactuals and applies this concept using neuro-fuzzy system and genetic algorithm for interstate conflict management. Furthermore, this book applies the theory of flexibly-bounded rationality for decision making by studying the problem of imperfect and incomplete information and its impact on decision making within the context of the theory of bounded rationality. To achieve this a multi-layer perceptron and particle swarm optimization are used in interstate conflict.

The theory of the marginalization of irrelevevant information is studied for decision making. In this regard four techniques are considered and these are the marginalization of irrationality approach, automatic relevance determination, principal component analysis and independent component analysis. These techniques are applied for condition monitoring, credit scoring, interstate conflict and face recognition.

Finally, this book studies the concept of group decision making and how artificial intelligence is used to facilitate decision making in a group. Four group based decision making techniques are considered and these are ensemble of support vector machines which are applied to land cover mapping incremental learning using genetic algorithm which is applied to optical character recognition, dynamically weighted mixtures of experts which are applied to platinum price prediction as well as the Learn++ which is applied to wine recognition.

This book is of value to students and professionals in the areas of engineering, politics and medicine.

Acknowledgement

I would like to thank the University of Johannesburg for contributing towards the writing of this book.

I also would like to thank my postgraduate students for their assistance in developing this manuscript. I thank Mr Reabetswe Nkhumise for assisting me with this manuscript.

I dedicate this book to the schools that gave me the foundation to always seek excellence in everything I do and these are: Mbilwi Secondary School, Case Western Reserve University, University of Pretoria, University of Cambridge (St. John' College) and Imperial College (London). I thank the Stellenbosch Institute of Advanced Study (STIAS) for financial support.

This book is dedicated to the following people: Nhlonipho Khathutshelo, Lwazi Thendo, Mbali Denga Marwala as well as Dr Jabulile Vuyiswa Manana.

Johannesburg   Professor Tshilidzi Marwala
1 October 2014

Contents

About the Author

Tshilidzi Marwala born in Venda (Limpopo, South Africa) is the Deputy Vice-Chancellor for Research at the University of Johannesburg. He was previously the Dean of Engineering at the University of Johannesburg, a Full Professor of Electrical Engineering, the Carl and Emily Fuchs Chair of Systems and Control Engineering as well as the South Africa Research Chair of Systems Engineering at the University of the Witwatersrand. He is a professor extraordinaire at the University of Pretoria and is on boards of EOH (Pty) Ltd and Denel. He is a Fellow of the following institutions: South African Academy of Engineering, South African Academy of Science, TWAS-The World Academy of Science and Mapungubwe Institute of Strategic Reflection. He is a senior member of the IEEE and distinguished member of the ACM. He is a trustee of the Bradlow Foundation as well as the Carl and Emily Fuchs Foundation. He is the youngest recipient of the Order of Mapungubwe and was awarded the President Award by the National Research Foundation. He holds a Bachelor of Science in Mechanical Engineering (*Magna Cum Laude*) from Case Western Reserve University, a Master of Engineering from the University of Pretoria, Ph.D. in Engineering from the University of Cambridge and completed a Program for Leadership Development at Harvard Business School. He was a post-doctoral research associate at Imperial College (London) and was a visiting fellow at Harvard University and Wolfson College at the University of Cambridge. His research interests include the applications of computational intelligence to engineering, computer science, finance, social science and medicine. He has successfully supervised 45 masters and 15 PhD students, published over 300 refereed papers and holds 3 patents. He is on the editorial board of the International Journal of Systems Science (Taylor and Francis) and his work has been featured in prestigious publications such as New Scientist. He has authored 6 books: Economic Modeling Using Artificial Intelligence Methods by Springer (2013); Condition Monitoring Using Computational Intelligence Methods by Springer (2012); Militarized Conflict Modeling Using Computational Intelligence Techniques by Springer (2011); Finite Element Model Updating Using Computational Intelligence Techniques by Springer (2010); Computational Intelligence for Missing Data Imputation, Estimation and Management: Knowledge Optimization Techniques by IGI Global Publications (2009); and Computational Intelligence for Modelling Complex Systems by Research India Publications (2007).

Chapter 1
Introduction to Rational Decision Making

1.1 Introduction

This book is essentially on rational decision making which is a complex process and has perplexed thinkers for a very long time. For example, on taking decisions rationally one ought to take into account that other players may be acting irrationally. In this situation, the rational decision that is taken will factor into account the irrational reaction from the other player. Suppose a doctor prescribes medicine to a superstitious individual. That decision making process (rational) of prescribing medicine might take into account of the individual's superstitious tendencies (irrational) and thus also prescribes that the patient should be supervised on taking the medication. So in essence the decision taken has factored into account of the fact that the patient will act irrationally.

In prehistoric society, decision making was jumbled with superstitions (Vyse 2000; Foster and Kokko 2009). For example there is an old superstitious belief that if a person encounters a black cat crossing the road then that person will encounter bad luck. The other superstition among the Venda people in South Africa is that if the owl crows in your house then there will be death in the family. How these superstitions came about is not known but one can speculate that perhaps sometimes in the past a powerful person in the community, perhaps the King, by chance encountered a black cat crossing the road and he had bad luck subsequently. Whatever the explanations for these superstitions may be, the fact remains that superstitions have been and remain part of who we are as human beings in this present state of our evolution. Perhaps sometimes in the future we will evolve into some other species which is not superstitious. Superstition is not limited to human beings only. Skinner (1948) conjectured that superstition has also been observed in other animals such as pigeons.

In this book superstition can be viewed as supernatural causality where something is caused by another without them linked to one another. The idea of one event causing another without any connection between them whatsoever is irrational.

© Springer International Publishing Switzerland 2014
T. Marwala, *Artificial Intelligence Techniques for Rational Decision Making*,
Advanced Information and Knowledge Processing,
DOI 10.1007/978-3-319-11424-8_1

Making decisions based on superstitious assumptions is irrational decision making and is the antithesis of this book which is on rational decision making.

This book basically proposes that the basic mechanisms of rational decision making are causality and correlation machines. Correlation is important because it fills in the gaps that exist in information needed for decision making because of the limitation of information in decision making. Furthermore, causality is important in decision making because it is through causal loops (if this then act so) that an appropriate cause of action is chosen from many possible causes of actions, based on the principle of maximizing the balance of utility, and thus reducing the complexity of the problem. Here we use the balance of utility because utility can be both positive and negative. Suppose we compare two ways of moving from A to B. The first choice is to use a car and the second one is to use a bicycle. The rational way is not just to select the choice which minimizes time but to compare this with the cost of the action. Rational decision making is a process of reaching a decision that maximizes the balance of utility (Habermas 1984) and these decisions are arrived at based on relevant information and by applying sound logic and optimizing resources (Etzioni 1988).

Nobel Prize Laureate Herbert Simon realized that on making a rational decision one does not always have all the information and the logic that is used is far from perfect and, consequently, he introduced the concept of bounded rationality (Simon 1991; Gigerenzer and Selten 2002). Marwala (2014) extended this notion by noticing that with the advent of artificial intelligence the bounds that Herbert Simon's theory prescribes are in fact flexible.

The next section explores what rational decision making is and why it is important.

1.2   What is Rational Decision Making?

It is important to first understand what rational decision making is and to do this it is important to understand the meaning of the words rational and decision making. According to the google dictionary, rational is defined as "based on or in accordance with reason or logic" while decision making is defined as "the action or process of making important decisions'. In this book we define rational decision making as a process of making decisions based on relevant information, in a logical, timely and optimized manner. Suppose a man called Rudzani wants to make a decision on how much coffee he should drink today and he calls his aunt Vho-Denga to find out the color of shoes she is wearing and uses this information to decide on the course of action on how much coffee he will drink on that day. This will be an irrational decision making process because Rudzani is using irrelevant information (the color of shoes his aunt Vho-Denga is wearing) to decide how much coffee he will drink. If on the same token Rudzani decides that every time he takes a sip of that coffee he needs to walk for 1 km, we then will conclude that he is acting irrationally because he is wasting energy unnecessarily by walking 1 km in order to take a sip of coffee.

Fig. 1.1 Steps for rational decision making

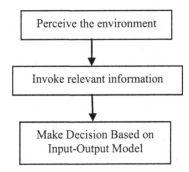

Fig. 1.2 A rational decision making framework

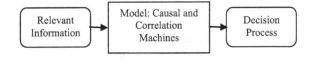

This unnecessary wastage of energy is irrational and in mathematical terminology we describe this as an un-optimized solution. If Rudzani decides that every time he sips the coffee he first pours it from one glass to another for no other reason except to fulfil the task of drinking coffee, then he is using an irrational course of action because this is not only illogical but also un-optimized.

An illustration of a rational decision making framework is illustrated in Fig. 1.1. This figure shows that on undertaking a rational decision making process one studies the environment or more technically a decision making space. Then the next step is to identify relevant information necessary for decision making. Then this information is presented to a decision engine which is logical and consistent. In this book, we propose that this decision engine is rational when it uses the correlation and causal machines to process relevant information in a logical, timely and optimized manner and this will be described in detail later. This decision engine then gives a course of action which should be followed to make a rational decision. The rational decision making process is illustrated in Fig. 1.2.

1.3 Flexibly-Bounded Rational Decision Making

In the previous section it was discussed that rationality comprises of the concept of making an optimized decision, in a logical and timely manner using relevant information. However, Simon (1991) observed that making an optimized decision is in most cases not practically possible, because information is incomplete and one does not have all the necessary intelligent infrastructure to make sense of such incomplete yet vast information. This is what is called the theory of bounded rationality and was first advanced by Nobel Laureate Herbert Simon. Bounded rationality can be viewed as an optimization problem where the objective function is not complete because not all the design variables can be observed and the objective function is thus but a rough approximation of the correct one. The consequence of the theory of bounded rationality on many practical and philosophical problems is substantial.

With advances in information processing techniques, enhanced theories of autoassociative machines and advances in artificial intelligence methods, the theory of bounded rationality was revised by Marwala (2014). The fact that information which is used to make decisions is imperfect because factors such as measurement errors can be partially corrected by using advanced information analysis methods and the fact that some of the data that are missing and, thereby, incomplete can be partially completed using missing data estimation methods, processing power is variable because of Moore's Law where computational limitations are advanced continuously (Frenzel 2014) and the fact that a human brain which is influenced by other social and physiological factors can be substituted for by recently developed artificial intelligence machines implies that the bounds under which rationality is exercised can be shifted and thus bounded rationality can now be transformed into the theory of flexibly-bounded rationality (Marwala 2014).

1.4 Correlation Machine

In Fig. 1.2, we indicated that the rational decision model is made up of correlation as well as causal machines. In this section, we describe the idea of a correlation machine. To demonstrate this concept of correlation, we study the relationship between walking every day and life expectancy. In this example, the two variables x indicating the aggregate hours a person has spent walking and y indicating life expectancy. If variable x is positively correlated to variable y then it means walking increases life expectancy or appropriately phrased the probability of a long life given the fact that a person walks is higher than one.

According to google dictionary correlation is defined as: "a mutual relationship or connection between two or more things." This relationship or connection is gauged by the degree to which variables vary together and can be quantified.

The concept of a correlation function was introduced by Marwala (2014). A correlation function is a predictive system that relates sets of variables that are only correlated and are not causal. A type of a correlation machine is the auto-associative network where a network is trained to recall itself (Kramer 1992). An auto-associative network is able to reconstruct a complete set of information from a small piece of information. For example, one could easily complete the following statement if one is familiar with the history of South Africa: "Nelson Mandela spent many years…"

Auto-associative networks have been used in a number of applications including in novelty detection, image compression (Rios and Kabuka 1995), missing data estimation (Marwala 2009), to maximize the number of correctly stored patterns (Masuda et al. 2012), in business failure prediction (Chen 2010), in gaze tracking (Proscevicius et al. 2010), for fault detection in power systems (Miranda et al. 2012)

Fig. 1.3 An auto-associative
network

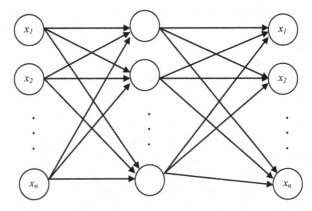

and to reconstruct color images corrupted by either impulsive or Gaussian noise
(Valle and Grande Vincente 2011). Fernandes et al. (2011) applied auto-associative
pyramidal neural network successfully for face verification whereas Jeong et al.
(2011) successfully applied auto-associative multilayer perceptron network for laser spot detection based on computer interface system. An illustration of the autoassociative network is shown in Fig. 1.3.

Figure 1.3 is a model which defines inter-relationships in the variables and the
rules that govern the data. The unknown variable(s) is identified by ensuring that
the correct estimated variable(s) are those that when identified conform to all the
inter-relationships that are present amongst the data and the rules that govern the
data. The unknown variable(s) is estimated by an optimization method that treats
the unknown variable(s) as design variable(s) and the interrelationships amongst the
data and rules that govern the data as the objective to be reached.

Of course the auto-associative network is not the only type correlation machine.
Another correlation machine is given the observations x and y one can construct an
equation $y = f(x)$ where y does not happen after x or there is no flow of information from x and y. For example, every time it is cloudy Khathu decides to carry an
umbrella. This will be because he has formulated a correlation machine in mind
which when x = cloudy, then y = carry an umbrella and x = not cloudy then y = do
not carry an umbrella. This will be a decision making system based on the correlation of cloud and rain.

Now that a correlation machine has been studied the next step is to study the
causal machine which is another element which is very important for rational decision making.

1.5   Causal Machine

As described before rational decision making entails the use of relevant information
and logic to make decisions in a timely manner. It was also described that the logical
aspect of rational decision making often manifests itself in the form of a causal and

correlation machines. This section describes the concept of a causal machine which is essentially an instrument that is able to give the outcome of a particular process given some observations.

In order to understand a causal machine it is imperative to understand the concept of causality. The English dictionary defines causality as "the relationship between cause and effect". The cause is defined as "making something", whereas the effect is "a change which is a result of an action". This book adopts the concept that for one variable to cause another there has to be a flow of information from the cause to the effect. There are many different types of causes and these are *prima facie* causes, spurious causes, direct causes, supplementary causes, sufficient causes and negative causes (Suppes 1970). B is the *prima facie* cause of A if B occurs before event A and the probability of B happening is greater than zero and the conditional probability of A given B is greater than the probability of A. Spurious cause which is also known as the confounding and is when A and B seems to show causal relations while in reality they are both caused by an independent variable C. In this instance the correlation machine will be appropriate to model the relationship between variables A and B. Direct cause is when variable A causes variable B without intermediary factors. In this instance a causal machine will be appropriate for modelling the relationship between variables A and B. Supplementary cause is when more than one cause supplements one another to cause an event. Sufficient causes are those causes that cause the event to definitely happen.

Many theories have been proposed to explain causality and some of them will be explained in Chap. 2. For example, Hume proposed the following principles of causality (Hume 1896; Marwala 2014):

- That the cause and effect are connected in space and time.
- That the effect should occur after the event.
- That the cause and effect should be continuously connected.
- That a particular cause always give identical effect and the same effect is always obtained from the same cause.
- If different causes give the same effect it should be because of what is common amongst them.
- The difference in the effects of two similar events must be because of the difference.

Granger (1969) summarized these principles to mean that the cause and effect are associated or correlated, that the cause happens before the effect and that the cause and effect are connected. These are the necessary but not sufficient conditions for causality.

There have been a number of theories of causality proposed and these will be explained in detail in the later chapters and these are the transmission theory of causality where causal information is transmitted from the cause to the effect (Ehring 1986; Kistler 1998; Salmon 1984; Dowe 1992). Another theory is the probability theory of causality which is based on the observation that some types of causality are probabilistic (Ten Kate 2010; Schmidtke and Krawczak 2010; Plotnitsky 2009).

Another theory of causality is the projectile theory of causality which was proposed by Marwala (2014) and is a generalized version of the transmission theory of causality where information is transmitted with a specific intensity and configuration like a projectile. The theory of causal calculus is derived from conditional probability theory and is premised on the idea of estimating interventional probability from conditional probabilities (Rebane and Pearl 1987; Verma and Pearl 1990; Pearl 2000; Ortigueira et al. 2012; Raginsky 2011; Pommereau 2004). Granger causality is a technique for revealing directed functional connectivity based on time series analysis of causes and effects (Croux and Reusens 2012).

The manipulation theory of causality considers causal relationships between causal variable x and effect variable y and considers changes in x called Δx and assess whether it leads to changes in y (Δy). If it does, then there is a causal relationship between x and y (Baedke 2012; Erisen and Erisen 2012; Silvanto and Pascual-Leone 2012; Mazlack 2011). If this is not the case but correlation exists between variables x and y, then there is another variable which both x and y depend on.

Process theory of causality considers the causal relationship between variable x and y and identifies the actual process of causality not its mirror (Anderson and Scott 2012; Rodrigo et al. 2011; Sia 2007; Samgin 2007). In counterfactual theory of causality, given a factual with an antecedent (cause) and a consequence (effect), the antecedent is altered and the new consequence is derived (Byrne 2005; Lewis 1973; Miller and Johnson-Laird 1976; Simon and Rescher 1966). Structural learning theory of causality is a technique of identifying connections between a set of variables. Structural learning is based on the premise that there are three causal substructures that define causal relationships between variables and these were identified by Wright (1921) and they are direct and indirect causation ($X{\rightarrow}Z{\rightarrow}Y$), common cause confounding ($X{\leftarrow}Z{\rightarrow}Y$) and a collider ($X{\rightarrow}Z{\leftarrow}Y$).

In this chapter we define a causality function which takes an input vector (x) which is the cause and propagate it into the effect (y). This can be appropriately represented mathematically as follows:

$$y = f(x) \qquad\qquad (1.1)$$

Here f is the functional mapping. This equation strictly implies that y is directly obtained from x. Of course this elegant equation is not strictly only applicable to the cause and effect but can still be valid if x and y are correlated. What this book argues is that a function philosophically and structurally assumes a causal model under certain conditions and these are that x and y are connected in space and time, that x must happen before y, as well as that there should be a flow of information between x and y. Flow of information implies connectivity and therefore the condition that eq. (1.1) is a causal function is that there should be a flow of information from x to y. If these conditions are not met then the function in eq. (1.1) merely becomes a correlation function. The next section describes artificial intelligence methods that are used in this book.

1.6   Introduction to Artificial Intelligence

Artificial Intelligence (AI) is a field which is about the construction of intelligent systems with the ability to learn and think (Hopfield 1982; Marwala and Lagazio 2011; Xing and Gao 2014). There are several kinds of AI systems and here we consider learning and optimization methods. Optimization based on artificial intelligence uses the contrivance of complex social organisms such as the flock of birds, colony of ants and school of fishes to design systems that are intelligent to adapt without human participation by representing dialectic relationships between individual and group intelligence. The next sections describe several artificial intelligence methods which are used in this book.

1.6.1   Neural Networks

In this book, neural networks are used to create models that are capable of transmitting information from the cause to the effect. This transmission process is enabled using neurons which transmit information from the cause to the effect. This in essence denotes that the effect depends on the cause. Furthermore, neural networks are also used to create a correlation machine.

There are many types of neural networks and these include the radial basis function (RBF) and the multi-layer perceptron (MLP) (Marwala 2009; Marwala 2013b). Neural networks have been applied effectively in numerous diverse areas of variable complexities including control of non-linear system (Chemachema 2012), traffic sign classification (Cireşan et al. 2012), dynamic portfolio optimization (Liu et al. 2012), adaptation of wireless sensor, actor networks (Cañete et al. 2012), water retention (Ebrahimi et al. 2014), estimation of solar radiation (Al-Shamisi et al. 2014) and in breast cancer diagnosis (Motalleb 2014).

The MLP is a feed-forward neural network model where the input variables are feed-forwarded until they become the effect. The MLP, therefore, estimates a relationship between sets of input data (cause) and a set of appropriate output (effect). It is based on the standard linear perceptron and makes use of three or more layers of neurons (nodes) with non-linear activation functions, and is more powerful than the perceptron. This is because it can distinguish data that is not linearly separable, or separable by a hyper-plane. The multi-layer perceptron has been used to model many complex systems in areas such as estimating invasive cerebrospinal fluid pressure (Golzan et al. 2012), monitoring electric load of residential buildings (Rababaah and Tebekaemi 2012), separating nonlinear source (Elmannai et al. 2012), monitoring gear dynamics (Sanz et al. 2012), engineering sciences (Marwala 2010), approximating the Vickers hardness of Mn-Ni-Cu-Mo austempered ductile iron (PourAsiabi et al. 2012), estimating wind speed (Culotta et al. 2012), automatically monitoring critical infrastructure (Marwala 2012), and forecasting odorant chemical class from descriptor values (Bachtiar et al. 2011).

The multi-layer perceptron neural network has multiple layers neural units, which are interconnected in a feed-forward configuration (Haykin 1999; Marwala 2013b). Each neuron in one layer is directly connected to the neurons of the subsequent layer. Two-layered multi-layer perceptron architecture is used in this book because of the universal approximation theorem, which states that a two layered architecture is adequate for multi-layer perceptron modelling (Haykin 1999; Marwala 2013a). The network can be described as follows by assuming a logistic function in the outer layer and a hyperbolic tangent function in the inner layer and is schematically shown in Fig. 1.3:

$$y_k = \frac{1}{1 + e^{-\left(\sum_{j=1}^{M} w_{kj}^{(2)} \tanh\left(\sum_{i=1}^{d} w_{ji}^{(1)} x_i + w_{j0}^{(1)}\right) + w_{k0}^{(2)}\right)}} \tag{1.2}$$

In eq. (1.2), $w_{ji}^{(1)}$ and $w_{ji}^{(2)}$ indicate weights in the first and second layer, respectively, going from input i to hidden unit j, M is the number of hidden units, d is the number of output units while $w_{j0}^{(1)}$ and $w_{k0}^{(2)}$ indicate the biases for the hidden unit j and the output unit k.

1.6.2 Fuzzy Sets

Fuzzy logic is a technique of mapping an input space onto an output space by means of a list of linguistic rules that consist of the if-then statements (Bih 2006). Fuzzy logic consists of four components: fuzzy sets, membership functions, fuzzy logic operators and fuzzy rules (Biacino and Gerla 2002). In conventional set theory, an object is either an element or is not an element of a specific set (Ferreirós 1999; Yahyaoui et al. 2014; Livi et al. 2014; Pinto 2014).

Consequently, it is possible to define if an object belongs to a specific set because a set has distinct boundaries, as long as an object cannot realize partial membership. The key objective of fuzzy logic is to permit a more flexible representation of sets of objects by using a fuzzy set. A fuzzy set does not have as clear cut boundaries as a classical set and the objects are characterized by a degree of membership to a specific set (Halpern 2003; Wright and Marwala 2006). Consequently, intermediate values of objects can be represented in a way that is closer to the way the human brain thinks, as opposed to the clear cut-off boundaries in classical sets.

1.6.3 Rough Sets

Rough set theory is a mathematical tool which deals with vagueness and uncertainty (Pawlak 1991; Zhang et al. 2014). It allows for the approximation of sets that are difficult to describe with available information. The advantages of rough sets, as with many other computational intelligence techniques, are that they do not require rigid a priori assumptions about the mathematical nature of such complex relationships, as commonly used multivariate statistical techniques do (Crossingham et al. 2008). Rough set theory is based on the assumption that the information of interest is associated with some information of its universe of discourse (Tettey et al. 2007; Crossingham et al. 2009; Li et al. 2014).

1.6.4 Hopfield Networks

Hopfield networks are memory systems with binary threshold nodes and can be trained to converge to an optimal point (Hopfield 1982). They are in essence models that allow for the comprehension of human memory. They are able to model associative memory and there are two categories of processes: auto-association and hetero-association. Auto-association is when a vector is associated with itself while hetero-association is when two different vectors are associated. Auto-association is used in this chapter to build a correlation function. Hopfield networks have been applied quite extensively for economic dispatch with multiple options of fuel (Dieu 2013), in optimal control (Wang and Nakagiri 2012), remote sensing (Santos et al. 2012; Sammouda et al. 2014), spectrum and power distribution (Yang and Jiang 2012), production planning (Su 2012), solving load balancing problems (Fei and Ding 2012), fault diagnosis (Li et al. 2012; Elhadef and Romdhane 2014) and in GPU platform (Mei et al. 2014). The concept of a Hopfield network is used to create an auto-associative network.

We also use neural networks to build an auto-associative memory network which is in essence a correlation machine because it is primarily based on correlations between the data (Hopfield 1982; Marwala 2009). It is called auto-associative memory network because it is able to reveal a complete set of information by observing a small piece of information. It is an associative memory network because learning is based on association of the small piece of information to the whole. In cognitive psychology association is a powerful tool which is useful for learning complex patterns.

1.6.5 Support Vector Machines

Support vector machines (SVMs) are supervised learning methods used mainly for classification, are derived from statistical learning theory (Vapnik 1995; Garšva and Danenas 2014; Chou et al. 2014). For SVMs, a data point is conceptualized as a p-dimensional vector. The aim is to separate such points with a p-1-dimensional hyperplane, known as a linear classifier. There are numerous hyperplanes that can be used. These include the one that exhibits the largest separation, also called the margin, between the two classes. The selected hyperplane can be chosen so that the distance from it to the nearest data point on both sides is maximized. This is then known as the *maximum-margin hyperplane*.

1.6.6   Genetic Algorithm

Genetic algorithm is a global optimization technique which is based on the principles of evolution as proposed by Darwin (1859). It consists of four motive forces and these are crossover where two members of a population mate to produce an offspring, mutation where a member of the population changes aspect of its genes and reproduces according to the principles of the survival of the fittest. Genetic algorithm was applied successfully to optimize resource levels in surgical services (Lin et al. 2013), optimize stay cables in bridges (Hassan 2013), optimize steel panels (Poirier et al. 2013), schedule shop floor (Wang and Liu 2013); task teams of unmanned aerial vehicles (Darrah et al. 2013), economic policy control (Razavi et al. 2014), in green transportation optimization (Lin et al. 2014) and for dispute classification (Chou et al. 2014). Genetic algorithm is used in this book to build a correlation function.

1.6.7   Particle Swarm Optimization

Particle swarm optimization technique is a stochastic, population-based evolutionary global optimization method that was motivated by algorithms that model the "flocking behavior" seen in birds (Kennedy and Eberhart 1995; Palma et al. 2014). Within the optimization framework, the notion of birds identifying a roost is similar to a process of identifying an optimal solution. Its main feature is its socio-psychological characteristics derived from swarm intelligence and offers comprehension of social behavior and thus it is governed by two drivers and these are group and individual knowledge. Particle swarm optimization method has been applied successfully to predict foreign exchange rates (Sermpinis et al. 2013), design bacterial foraging in power system stabilizers (Abd-Elazim and Ali 2013), model elasto-plastic of unsaturated soils (Zhang et al. 2013a), plan robot paths (Zhang et al. 2013b) and design electronic enclosure (Scriven et al. 2013; Garšva and Danenas 2014; Palma et al. 2014). Particle Swarm Optimization is used to model a correlation machine in this book.

1.6.8   Simulated Annealing

Simulated annealing is a probabilistic meta-heuristic global optimization technique that is based on natural annealing process (Kirkpatrick et al. 1983; He et al. 2014; Borges et al. 2014). Critical elements of simulated annealing are the energy function which is a mathematical or heuristical representation of the objective function and the temperature. On implementing simulated annealing the temperature is reduced gradually and this is called temperature schedule until it is frozen when the temperature is equal to 1. Simulated annealing has found many successful applications such as identification of order picking policy (Atmaca and Ozturk 2013), vehicle routing (Baños et al. 2013), design of pre-stressed concrete precast road bridges (Martí et al. 2013), inferring maximum parsimony phylogenetic (Richer et al. 2013), mixed model assembly (Manavizadeh et al. 2013) and for optimal planning (Popović et al. 2014). Simulated annealing is used to model a correlation machine in this book.

1.7 Summary and Outline of the Book

Rational decision making, flexibly-bounded rationality, artificial intelligence, causality and correlation were discussed in this chapter. In Chap. 2, a causal machine is defined and built using various artificial intelligence techniques whereas in Chap. 3, a correlation machine is defined and built using various artificial intelligence techniques. Chap. 4 applies the correlation and causal machines for rational decision making. Chap. 5 applies the theory of counterfactuals for rational decision making. Chap. 6 applies the theory of flexibly-bounded rationality for decision making by studying the problem of imperfect and incomplete information and its impact on decision making within the context of the theory of bounded rationality. Chap. 7 studies methods for dealing with irrelevant information in decision making. Chap. 8 applies artificial intelligence for rational group decision making. This book is then concluded in Chap 9.

Conclusions

This chapter discussed flexibly-bounded rationality, rational decision making causality, correlation and artificial intelligence. A correlation function was used to build a correlation machine. Causality was described and applied to build a causal machine. The artificial intelligence methods presented in this chapter included neural networks, support vector machines, rough sets, fuzzy system, particle swarm optimization, genetic algorithm and simulated annealing.

References

Abd-Elazim SM, Ali ES (2013) A hybrid Particle swarm optimization and bacterial foraging for optimal power system stabilizers design. Int J Electr Power Energy Syst 46(1):334–341
Al-Shamisi M, Assi A, Hejase H (2014) Estimation of global solar radiation using artificial neural networks in Abu Dhabi city, United Arab Emirates. J Sol Energy Eng Trans ASME 136(2), art. no. 024502

Anderson GL, Scott J (2012) Toward an intersectional understanding of process causality and social context. Qual Inq 18(8):674–685

Atmaca E, Ozturk A (2013) Defining order picking policy: a storage assignment model and a simulated annealing solution in AS/RS systems. Appl Math Model 37(7):5069–5079

Bachtiar LR, Unsworth CP, Newcomb RD, Crampin EJ (2011) Predicting odorant chemical class from odorant descriptor values with an assembly of multi-layer perceptrons. Proceedings of the Annual International Conference of the IEEE Engineering in Medicine and Biology Society, EMBS, art. no. 6090755, pp 2756–2759

Baedke J (2012) Causal explanation beyond the gene: manipulation and causality in epigenetics. Theoria (Spain) 27(2):153–174

Baños R, Ortega J, Gil C, Fernández A, De Toro F (2013) A Simulated Annealing-based parallel multi-objective approach to vehicle routing problems with time windows. Expert Syst Appl 40(5):1696–1707

Biacino L, Gerla G (2002) Fuzzy logic, continuity and effectiveness. Arch Math Logic 41:643–667

Bih J (2006) Paradigm shift—an introduction to fuzzy logic. IEEE Potentials, vol. 25, no. 1. New Jersey, pp 6–21

Borges P, Eid T, Bergseng E (2014) Applying simulated annealing using different methods for the neighborhood search in forest planning problems. Eur J Oper Res 233(3):700–710

Byrne RMJ (2005) The rational imagination: how people create counterfactual alternatives to reality. MIT Press, Cambridge

Cañete E, Chen J, Marcos Luque R, Rubio B (2012) NeuralSens: a neural network based framework to allow dynamic adaptation in wireless sensor and actor networks. J Netw Comput Appl 35(1):382–393

Chemachema M (2012) Output feedback direct adaptive neural network control for uncertain SISO nonlinear systems using a fuzzy estimator of the control error. Neural Netw 36:25–34

Chen M-H (2010) Pattern recognition of business failure by autoassociative neural networks in considering the missing values. ICS 2010—International Computer Symposium, art. no. 5685421, pp 711–715

Chou J-S, Cheng M-Y, Wu Y-W, Pham A-D (2014) Optimizing parameters of support vector machine using fast messy genetic algorithm for dispute classification. Expert Syst Appl 41(8):3955–3964

Cireşan D, Meier U, Masci J, Schmidhuber J (2012) Multi-column deep neural network for traffic sign classification. Neural Netw 32:333–338

Crossingham B, Marwala T, Lagazio M (2008) Optimised rough sets for modeling interstate conflict. In: Proceeding of the IEEE International Conference on Systems, Man, and Cybernetics, pp 1198–1204

Crossingham B, Marwala T, Lagazio M (2009) Evolutionarily optimized rough set partitions. ICIC Express Lett 3:241–246

Croux C, Reusens P (2012) Do stock prices contain predictive power for the future economic activity? A Granger causality analysis in the frequency domain, Journal of Macroeconomics, Article in Press

Culotta S, Messineo A, Messineo S (2012) The application of different model of multi-layer perceptrons in the estimation of wind speed. Adv Mater Res 452–453:690–694

Darrah M, Fuller E, Munasinghe T, Duling K, Gautam M, Wathen M (2013) Using genetic algorithms for tasking teams of raven UAVs. J Intell Robot Syst: Theory App 70(1–4):361–371

Darwin C (1859) On the origin of species by means of natural selection, or the preservation of favoured races in the struggle for life. John Murray, London

Dieu VN, Ongsakul W, Polprasert J (2013) The augmented Lagrange Hopfield network for economic dispatch with multiple fuel options. Math Comp Model 57(1–2):30–39

Dowe TP (1992) Wesley Salmon's process theory of causality and conserved quantity theory. Philos Sci 59:195–216

Ebrahimi E, Bayat H, Neyshaburi MR, Zare Abyaneh H (2014) Prediction capability of different soil water retention curve models using artificial neural networks. Arch Agron Soil Sci 60(6):859–879

Ehring D (1986) The transference theory of causation. Synthese 67:249–258

Elhadef M, Romdhane LB (2014) Fault diagnosis using partial syndromes: a modified Hopfield neural network approach. Int J Parallel Emerg Distrib Syst 29(2):119–146

Elmannai H, Loghmari MA, Karray E, Naceur MS (2012) Nonlinear source separation based on Multi-Layer Perceptron: application on remote sensing analysis. 2012 2nd International Conference on Remote Sensing, Environment and Transportation Engineering, RSETE 2012–Proceedings, art. no. 6260640

Erisen E, Erisen C (2012) The effect of social networks on the quality of political thinking. Polit Psychol 33(6):839–865

Etzioni A (1988) Normative-affective factors: towards a new decision-making model. J Econ Psychol 9:125–150

Fei C, Ding F (2012) Solving the load balancing problem by adding decaying continuous Hopfield neural network. 2012 2nd international conference on applied robotics for the power industry, CARPI 2012, art. no. 6356299, pp 276–279

Fernandes BJT, Cavalcanti GDC, Ren TI (2011) Autoassociative pyramidal neural network for face verification. Proceedings of the international joint conference on neural networks, art. no. 6033417, pp 1612–1617

Ferreirós J (1999) Labyrinth of thought: a history of set theory and its role in modern mathematics. Birkhäuser, Basel

Foster KR, Kokko H (2009) The evolution of superstitious and superstition-like behaviour. Proc Royal Soc B: Biol Sci 276(1654):31–7

Frenzel L (2014) Is Moore's law really over for good? Electron Des 62(3):5–8

Garšva G, Danenas P (2014) Particle swarm optimization for linear support vector machines based classifier selection. Nonlinear Anal: Model Control 19(1):26–42

Gigerenzer G, Selten R (2002) Bounded rationality: the adaptive toolbox, MIT Press, Cambridge

Golzan SM, Avolio A, Graham SL (2012) Non-invasive cerebrospinal fluid pressure estimation using multi-layer perceptron neural networks. Proceedings of the Annual International Conference of the IEEE Engineering in Medicine and Biology Society, EMBS, art. no. 6347185, pp 5278–5281

Granger CWJ (1969) Investigating causal relations by econometric models and cross-spectral methods. Econometrica 37:424–438

Habermas J (1984) The theory of communicative action volume 1; reason and the rationalization of society. Polity Press, Cambridge

Halpern JY (2003) Reasoning about uncertainty. Mass: MIT Press, Cambridge

Hassan MM (2013) Optimization of stay cables in cable-stayed bridges using finite element, genetic algorithm, and B-spline combined technique. Eng Struct 49:643–654

Haykin S (1999) Neural networks: a comprehensive foundation, 2nd edn, Prentice-Hall, New Jersey

He J, Dai H, Song X (2014) The combination stretching function technique with simulated annealing algorithm for global optimization. Optim Methods Softw 29(3):629–645

Hopfield JJ (1982) Neural networks and physical systems with emergent collective computational abilities. Proc Natl Acad Sci USA 79(8):2554–2558

Hume D, Selby-Bigge LA (ed) (1896) A treatise of human nature, Clarendon Press, Oxford

Jeong S, Jung C, Kim C-S, Shim JH, Lee M (2011) Laser spot detection-based computer interface system using autoassociative multilayer perceptron with input-to-output mapping-sensitive error back propagation learning algorithm. Opt Eng 50(8):1–11, art. no. 084302

Kennedy J, Eberhart RC (1995) Particle swarm optimization. presented at the Proceedings of the IEEE International Joint Conference on Neural Networks

Kirkpatrick S, Gelatt CD, Vecchi MP (1983) Optimization by simulated annealing. Science 220(4598):671–680

Kistler M (1998) Reducing causality to transmission. Erkenntnis 48:1–24

Kramer MA (1992) Autoassociative neural networks. Comput Chem Eng 16(4):313–328

Lewis D (1973) Counterfactuals. Blackwell Publishers, New Jersey

Li C, Yang Y, Jia M, Zhang Y, Yu X, Wang C (2014) Phylogenetic analysis of DNA sequences based on k-word and rough set theory. Physica A: Stat Mec Appl 398:162–171

Li P, Xiong Q, Chai Y, Wang K (2012) Analog fault diagnosis using hopfield network and multiscale wavelet transform method. J Comput Inf Syst 8(23):9721–9728

Lin C, Choy KL, Ho GTS, Ng TW (2014) A Genetic Algorithm-based optimization model for supporting green transportation operations. Expert Syst Appl 41(7):3284–3296

Lin R-C, Sir MY, Pasupathy KS (2013) Multi-objective simulation optimization using data envelopment analysis and genetic algorithm: specific application to determining optimal resource levels in surgical services. Omega (United Kingdom) 41(5):881–892

Liu Q, Guo Z, Wang J (2012) A one-layer recurrent neural network for constrained pseudoconvex optimization and its application for dynamic portfolio optimization. Neural Netw 26:99–109

Livi L, Tahayori H, Sadeghian A, Rizzi A (2014) Distinguishability of interval type-2 fuzzy sets data by analyzing upper and lower membership functions. Appl Soft Comput J 17:79–89.

Manavizadeh N, Hosseini N-S, Rabbani M, Jolai F (2013) A simulated annealing algorithm for a mixed model assembly U-line balancing type-I problem considering human efficiency and Just-In-Time approach. Comp Ind Eng 64(2):669–685

Martí JV, Gonzalez-Vidosa F, Yepes V, Alcalá J (2013) Design of prestressed concrete precast road bridges with hybrid simulated annealing. Eng Struct 48:342–352

Marwala T (2009) Computational intelligence for missing data imputation, estimation and management: knowledge optimization techniques, Information Science Reference Imprint. IGI Global Publications, New York

Marwala T (2010) Finite element model updating using computational intelligence techniques, Springer-Verlag, London

Marwala T (2012) Condition monitoring using computational intelligence methods. Springer-Verlag, London

Marwala T (2013a) Causality, correlation and artificial intelligence: implication on policy formulation. The Thinker 49:36–37

Marwala T (2013b) Economic modelling using artificial intelligence methods. Springer-Verlag, London. (in press)

Marwala T (2014) Causality, correlation and artificial intelligence for rational decision making. World Scientific Publications, Singapore

Marwala T, Lagazio M (2011) Militarized conflict modeling using computational intelligence techniques. Springer-Verlag, London

Masuda K, Fukui B, Kurihara K (2012) A weighting approach for autoassociative memories to maximize the number of correctly stored patterns. Proceedings of the SICE annual conference, art. no. 6318692, pp 1520–1524

Mazlack LJ (2011) Approximate computational intelligence models and causality in bioinformatics. Proceedings of the 6th IASTED international conference on computational intelligence and bioinformatics, CIB 2011, pp 1–8

Mei S, He M, Shen Z (2014) Optimizing Hopfield neural network for spectral mixture unmixing on GPU platform. IEEE Geosci Remote Sens Lett 11(4):818–822, art. no. 6623088

Miller G, Johnson-Laird PN (1976) Language and perception. Cambridge University Press, Cambridge

Miranda V, Castro ARG, Lima S (2012) Diagnosing faults in power transformers with autoassociative neural networks and mean shift. IEEE Trans Power Deliv 27(3):1350–1357, art. no. 6176288

Motalleb G (2014) Artificial neural network analysis in preclinical breast cancer. Cell Journal 15(4):324–331

Ortigueira MD, Rivero M, Trujillo JJ (2012) The incremental ratio based causal fractional calculus. Int J Bifurcat Chaos 22(4):1–9

Palma G, Bia P, Mescia L, Yano T, Nazabal V, Taguchi J, Moréac A, Prudenzano F (2014) Design of fiber coupled Er3 + : chalcogenide microsphere amplifier via particle swarm optimization algorithm. Opt Eng 53(7):1–8, art. no. 071805

Pawlak Z (1991) Rough sets—theoretical aspects of reasoning about data. Kluwer Academic Publishers, Dordrecht

Pearl J (2000) Causality: models, reasoning, and inference. Cambridge University Press, Cambridge

Pinto A (2014) QRAM a qualitative occupational safety risk assessment model for the construction industry that incorporate uncertainties by the use of fuzzy sets. Safety Sci 63:57–76.

Plotnitsky A (2009) Causality and probability in quantum mechanics. AIP Conf Proc 1101:150–160

Poirier JD, Vel SS, Caccese V (2013) Multi-objective optimization of laser-welded steel sandwich panels for static loads using a genetic algorithm. Eng Struct 49:508–524

Pommereau F (2004) Causal time calculus. Lect Notes Comput Sci (including subseries Lecture Notes in Artificial Intelligence and Lecture Notes in Bioinformatics) 2791:260–272

Popović ZN, Kerleta VD, Popović DS (2014) Hybrid simulated annealing and mixed integer linear programming algorithm for optimal planning of radial distribution networks with distributed generation. Electr Pow Syst Res 108:211–222

PourAsiabi H, PourAsiabi H, AmirZadeh Z, BabaZadeh M (2012) Development a multi-layer perceptron artificial neural network model to estimate the Vickers hardness of Mn-Ni-Cu-Mo austempered ductile iron. Mater Design 35:782–789

Proscevicius T, Raudonis V, Kairys A, Lipnickas A, Simutis R (2010) Autoassociative gaze tracking system based on artificial intelligence. Elektronika ir Elektrotechnika 5:67–72

Rababaah AR, Tebekaemi E (2012) Electric load monitoring of residential buildings using goodness of fit and multi-layer perceptron neural networks. CSAE 2012– Proceedings, 2012 IEEE international conference on computer science and automation engineering. 2, art. no. 6272871, pp 733–737

Raginsky M (2011) Directed information and Pearl's causal calculus. 2011 49th annual allerton conference on communication, control, and computing, allerton 2011. Art. No. 6120270, pp 958–965

Razavi H, Ramezanifar E, Bagherzadeh J (2014) An economic policy for noise control in industry using genetic algorithm. Safety Sci 65:79–85

Rebane G, Pearl J (1987) The recovery of causal poly-trees from statistical data. Proceedings, 3rd workshop on uncertainty in AI, (Seattle, WA), pp 222–228

Richer J-M, Rodriguez-Tello E, Vazquez-Ortiz KE (2013) Maximum parsimony phylogenetic inference using simulated annealing. Adv Intell Syst Comput 175:189–203 ADVANCES

Rios A, Kabuka M (1995) Image compression with a dynamic autoassociative neural network. Math Comput Model 21(1–2):159–171

Rodrigo M, Liberos A, Guillem MS, Millet J, Climent AM (2011) Causality relation map: a novel methodology for the identification of hierarchical fibrillatory processes. Comput Cardiol 38:173–176. art. no. 6164530

Salmon W (1984) Scientific explanation and the causal structure of the world. Princeton University Press, Princeton

Samgin AL (2007) On an application of the causality principle to the theory of ion transport processes. J Phys Chem Solids 68(8):1561–1565

Sammouda R, Adgaba N, Touir A, Al-Ghamdi A (2014) Agriculture satellite image segmentation using a modified artificial Hopfield neural network. Comput Hum Behav 30:436–441

Santos S, Castaneda R, Yanez I (2012) Hopfield and pulse coupled neural networks aggregation for enhanced change detection in remote sensing imagery. World Automation Congress Proceedings, art. no. 6320941

Sanz J, Perera R, Huerta C (2012) "Gear dynamics monitoring using discrete wavelet transformation and multi-layer perceptron neural networks". Appl Soft Comput J 12(9):2867–2878

Schmidtke J, Krawczak M (2010) Psychomotor developmental delay and epilepsy in an offspring of father-daughter incest: quantification of the causality probability. Int J Legal Med 124(5):449–450

Scriven I, Lu J, Lewis A (2013) Electronic enclosure design using distributed particle swarm optimization. Eng Optimiz 45(2):167–183

Sermpinis G, Theofilatos K, Karathanasopoulos A, Georgopoulos EF, Dunis C (2013) Forecasting foreign exchange rates with adaptive neural networks using radial-basis functions and Particle Swarm Optimization. Eur J Oper Res 225(3):528–540

Sia S (2007) Creative synthesis: a process interpretation of causality. Philosophia 36(2):213–221

Silvanto J, Pascual-Leone A (2012) Why the assessment of causality in brain-behavior relations requires brain stimulation. J Cognitive Neurosci 24(4):775–777

Simon H (1991) Bounded rationality and organizational learning. Organ Sci 2(1):125–134

Simon H, Rescher N (1966) Cause and counterfactual. Philos Sci 33:323–340

Skinner BF (1948) Superstition' in the Pigeon. J Exp Psychol 38(2):168–172

Su J-L (2012) Intelligence decision supporting algorithms of production planning based on Hopfield network. Proceedings of the world congress on intelligent control and automation (WCICA), art. no. 6359166, pp 4122–4125

Suppes P (1970) A probabilistic theory of causality. North-Holland Publishing Company, Amsterdam

Ten Kate LP (2010) Psychomotor developmental delay and epilepsy in an offspring of father-daughter incest: quantification of the causality probability. Int J Legal Med 124(6):667–668

Tettey T, Nelwamondo FV, Marwala T (2007) HIV data analysis via rule extraction using rough sets. Proc of the 11th IEEE International Conference on Intelligent Engineering Systems, pp 105–110

Valle ME, Grande Vicente DM (2011) Some experimental results on sparsely connected autoassociative morphological memories for the reconstruction of color images corrupted by either impulsive or Gaussian noise. Proceedings of the International Joint Conference on Neural Networks, art. no. 6033232, pp 275–282

Vapnik VN (1995) The nature of statistical learning theory. Springer-Verlag. Berlin, Heidelberg

Verma T, Pearl J (1990) Equivalence and synthesis of causal models. Proceedings of the sixth conference on uncertainty in artificial intelligence, (July, Cambridge, MA), pp 220–227

Vyse SA (2000) Believing in magic: the psychology of superstition. Oxford University Press, Oxford

Wang Q-F, Nakagiri S-I (2012) Sensitivity of optimal control for diffusion Hopfield neural network in the presence of perturbation. Appl Math Comput 219(8):3793–3808

Wang S, Liu M (2013) A genetic algorithm for two-stage no-wait hybrid flow shop scheduling problem. Comput Oper Res 40(4):1064–1075

Wright S (1921) Correlation and causation. J Agric Res 7(3):557–585

Wright S, Marwala T (2006) Artificial intelligence techniques for steam generator modelling. arXiv:0811.1711

Xing B, Gao W-J (2014) Innovative computational intelligence: a rough guide to 134 clever algorithms. Springer-Verlag, London

Yahyaoui H, Almulla M, Own HS (2014) A novel non-functional matchmaking approach between fuzzy user queries and real world web services based on rough sets. Future Gener Comp Syst 35:27–38.

Yang M, Jiang M (2012) Hybrid spectrum access and power allocation based on improved hopfield neural networks. Adv Mater Res 588–589:1490–1494

Zhang M, Xu W, Yang X, Tang Z (2014) Incomplete variable multigranulation rough sets decision. Appl Math Inf Sci 8(3):1159–1166

Zhang Y, Gallipoli D, Augarde C (2013a) Parameter identification for elasto-plastic modelling of unsaturated soils from pressuremeter tests by parallel modified particle swarm optimization. Comput Geotech 48:293–303

Zhang Y, Gong D-W, Zhang J-H (2013b) Robot path planning in uncertain environment using multi-objective particle swarm optimization. Neurocomputing 103:172–185

Chapter 2
Causal Function for Rational Decision Making: Application to Militarized Interstate Dispute

This chapter describes and defines a causal function within the context of rational decision making. This is implemented using rough sets to build the causal machines. The rough sets were successfully used to identify the causal relationship between the militarized interstate dispute variables (causes) and conflict status effects.

2.1 Introduction

Rational decision decision making is the process of making decisions based on logic. In essence logic here refers to using relevant information and choosing a course of action which minimizes the energy required to execute a task. The concept of nature prefering the path of least resistance is a natural phenomenon which has been studied extensively throughout history. Examples of the applications of the principle of the path of least resistance are included in Calculus of Variation (Gelfand and Fomin 2000; van Brunt 2004; Ferguson 2004; Bertsekas 2005; Bellman 1954). Basically in rational decision making it does not matter what the objective is but rather that to make a decision rational, it matters how one arrives at the objective. In this regard some of the most atrocious events in histroy such as genocides and slavery might be classified as having being executed rationally as long as on reaching a goal relevant infomation was used and the process followed was logical including the observation of the principles of least resistance and the minimum energy.

There is a game where there are two parents with a daughter and son who need to cross a wide river (Anonymous 2012). They can only get to the other side by borrowing a boat from a fisherman. Nevertheless, the boat can only carry one adult or two children at a time. What is the rational way in which a family can cross the river and return the boat to the fisherman? The rational way of doing this is to cross from this side A to the other side B using the following steps:

1. First the children cross the river to side B.
2. Then the son returns to side A.
3. Then the father crosses to side B to join his daughter.

© Springer International Publishing Switzerland 2014
T. Marwala, *Artificial Intelligence Techniques for Rational Decision Making,*
Advanced Information and Knowledge Processing,
DOI 10.1007/978-3-319-11424-8_2

4. Then the daughter returns to side A to pick her brother up and they both go to side B to join the father.
5. Then the son returns to side A to give the boat to the mother who crosses to side B to join the father and daughter.
6. Then the daughter gets into the boat and goes to her brother in side A and they both return to their parents in side B.
7. Then the daughter gets off in side B and the son returns to side A to give the boat to the fisherman.
8. Then the fisherman and the son go to side B to drop the son.
9. Then the fisherman returns to side A.

The boat crossed the river 13 times and this can be represented in the state space as in Table 2.1.

The process followed to achieve a goal was rational because it used relevant information that only the parents cannot be in the boat at the same time otherwise it will sink and that both children or a child and a fisherman can be on the boat at the same time. If this relevant information was not used (acting irrationally) then the boat could have sunk. Secondly, the whole process took 13 trips and could not have been done using shorter number of trips. If, however, this trip was done in more trips then this would have been irrational because the principle of minimum energy (path of least resistance) would have been violated. The process of decision making based on relevant information which maximizes the balance of utility is a rational decision. We use balance of utility because utility can be both positive and negative. Suppose we compare two ways of moving from A to B. The first action is to use a car and the second is to use a bicycle. The rational way is not just to choose a way

Table 2.1 State space of the distribution of the parents, children, fisherman and the boat. Here *bold* represents the location of the boat while F stands for the father, M for the mother, S for the son, D for the daughter and F_i for the fisherman

State	Side A	Side B
1	**FMSDF$_i$**	
2	**FMF$_i$**	SD
3	**FMSF$_i$**	D
4	**MSF$_i$**	DF
5	**MDSF$_i$**	F
6	**MF$_i$**	FDS
7	**MSF$_i$**	FD
8	**SF$_i$**	MFD
9	**DSF$_i$**	MF
10	**F$_i$**	DSMF
11	**F$_i$S**	DMF
12		**FMSDF$_i$**
13	**F$_i$**	SDMF

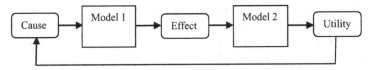

Fig. 2.1 Rational decision making process that uses causal functions

which minimizes time but to compare this with the cost of the choice made. In order to achieve maximum balance of utility the principle of minimum energy should be observed on reaching a decision and we term this a logical way of reaching a decision. In this chapter we introduce the causal machine for rational decision making (Marwala 2014).

The rational decision making process that uses causal functions is illustrated in Fig. 2.1. Here the causal function (Model 1) which links the cause to the effect is identified. The effect is linked to a utility by another model (Model 2). Rational decision making means identifying the cause that maximizes the balance of utility. The framework in Fig. 2.1 is collectively known as a causal machine. This chapter deals with a causal function (Model 1) and several models are explored in this regard.

In this chapter the model which is implemented uses the neuro-rough set to identify the causal relationship between demographic characteristics and HIV risk (Simiński 2012; Marwala and Crossingham 2008; Marwala and Crossingham 2009; Shamsuddin et al. 2009; Sabzevari and Montazer 2008; Choudhari et al. 2005; Czyzewski and Królikowski 2001).

The next section defines causality which is the basis of a causal machine.

2.2 Defining Causality

In this book it is proposed that causality is one of the elements that can be used for rational decision making. An example of causality is illustrated in Fig. 2.2. This figure illustrates three states of two balls. In State 1 a white ball is pushed to move towards a black ball. In State 2, the white ball hits the black ball. In State 3, both the white and the black balls are moving. It is clear here that the cause of the black

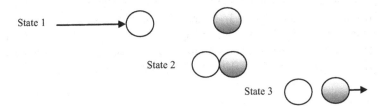

Fig. 2.2 Illustration of causality using two balls

ball moving is it being struck by the white ball. It is also clear that there was a transmission of information from the white ball to the black ball and that information is energy. As explained in Chap. 1 for event A to cause event B, there should be a flow of information from A to B, and A should happen before B. It is clear that events in Fig. 2.2 illustrate causality i.e. exchange of information (energy in this case) and that the cause happened before the effect. The relationship between the cause and the effect can be quantified easily using Newtonian Mechanics and the principle conservation of energy.

2.3   Models of Causality

In order to create a causal machine it is important to study different theories that have been proposed to explain causality. This section studies different models of causality as have previously been conceptualized and these are the transmission, probability, projectile, causal calculus, manipulation, process, counterfactual and structural theories of causality (Marwala 2014).

2.3.1   Transmission Theory of Causality

The transmission theory of causality is a model where causal information is transmitted from the cause to the effect (Sokolovski and Akhmatskaya 2012; Liu et al. 2012; Hubbard 2012; Ehring 1986; Kistler 1998; Salmon 1984; Dowe 1992). The information which is transmitted depends on the specific case being analyzed and in Fig. 2.2 it is energy.

2.3.2   Probability Theory of Causality

The probability theory of causality is based on the fact that some classes of causality are probabilistic (Mitroff and Silvers 2013; Ramsahai 2013; Kim 2013). For example, there is a conventional statement that *smoking causes lung cancer* (Yan et al. 2014; Park et al. 2014). There is a correlation between Smoking and Lung Cancer, but not all people who smoke get lung cancer. In the deterministic world smoking does not cause cancer. Nevertheless, in the probabilistic world smoking most definitely does cause lung cancer. To be specific, the phrase should be expressed as follows: the probability of lung cancer being present given that the subject smokes is high but is not necessarily definite. Consequently, a generalized causal model is necessarily probabilistic with the deterministic version being a special case.

2.3.3 Projectile Theory of Causality

The projectile theory of causality is a generalized version of the transmission theory of causality (Marwala 2014). The projectile theory of causality assumes that some information is transmitted from the cause to the effect in a form of a projectile with a specific intensity and configuration. The information that is transmitted from the cause to the effect can leave the point of cause with a certain velocity and sometimes reach the point of effect or sometimes falls short of the target and from this it is possible to build a probability distribution of causation.

2.3.4 Causal Calculus

Causal calculus is a model of inferring interventional probabilities from conditional probabilities (Rebane and Pearl 1987; Verma and Pearl 1990; Pearl 2000). Suppose we would like to estimate the hypothesis that HIV causes AIDS then causal calculus permits us to estimate the interventional probability that a person who is forced to have HIV (obviously illegal) develops AIDS $P(AIDS|forced(HIV))$ from the conditional probability that a person develops AIDS given the fact that he/she has HIV $P(AIDS|HIV)$. Causal calculus assumes that the structure that connects variables is in existence and where the structure does not exist it can be identified from the observed data using Bayesian networks.

2.3.5 Manipulation Theory of Causality

The manipulation theory of causality considers causal relationships between a causal variable x and an effect variable y and considers changes in x called Δx and assesses whether it leads to changes in y (Δy) in the model $y=f(x)$ (Baedke 2012; Annus et al. 2008; Hausman and Woodward 2004). If it does, then there is a causal relationship between x and y. If this is not the case then there is another variable which both x and y depend on.

2.3.6 Process Theory of Causality

Process theory of causality considers the causal relationship between variable x and y and identifies the actual process of causality not its mirror (Dowe 1992; Salmon 1998). The difficulty with this model lies in differentiating between the actual cause and effect from its mirror.

2.3.7 *Counterfactual Theory*

In counterfactual thinking, given a factual with an antecedent (cause) and a consequence (effect), the antecedent is altered and the new consequence is derived (Maudlin 2004). Suppose one would like to test what happens to the sugar level in a patient when insulin is administered. Then one would observe what happens to the sugar level when nothing is administered and compare this to when insulin is administered. In this instance not administering insulin is called a factual while administering insulin is called a counterfactual. Ideally for the efficacy of insulin to be observed then not administering insulin (factual) and administering insulin (counterfactual) should happen at the same time which is physically impossible. The compromise is for the factual and counterfactual to happen at nearly identical conditions (Collins et al. 2004; Loewer 2007).

2.3.8 *Structural Learning*

Structural learning identifies connections between a set of variables. In structural learning there are three causal substructures that define relationships between variables (Wright 1921) and these are direct and indirect causation ($X \rightarrow Z \rightarrow Y$), common cause confounding ($X \leftarrow Z \rightarrow Y$) and a collider ($X \rightarrow Z \leftarrow Y$). In structural learning relationships between variables are identified using heuristic optimization techniques. Now that we have discussed different causal models, the next section describes the causal function.

2.4   Causal Function

In this section we define a causal function which is a function that takes an input vector (x) and propagate it into the effect (y) where y happens after x and there is a flow of information between x and y. This function can be appropriately represented mathematically as follows:

$$y = f(x) \tag{2.1}$$

Here f is the functional mapping. This equation strictly implies that y is directly obtained from x. Of course this elegant equation is not strictly only applicable to the cause and effect but can still be valid if x and y are correlated and thus become a correlation function if either or both of the conditions (1) y happens after x, and (2) there is a flow of information from x to y are violated.

To illustrate the concept of a causal function we will use a classical problem in Physics of a ball colliding with a stationary ball which is illustrated in Fig. 2.2. To derive the causal function which relates the character of the cause to the character of the effect we apply the principles of conservation of energy and momentum assuming that the collision is perfectly elastic. The conservation of energy can be written as follows:

$$\frac{1}{2}m_1u_1^2 = \frac{1}{2}m_1v_1^2 + \frac{1}{2}m_2v_2^2 \Rightarrow m_1u_1^2 = m_1v_1^2 + m_2v_2^2 \tag{2.2}$$

Here m_1 and m_2 are masses of the white and black balls, respectively while u_1 and u_2 (here u_2 is equal to zero) are the velocities of white and black balls respectively before collision whereas v_1 and v_2 are the velocities of white and black balls respectively after collision. Using the principle conservation of momentum we obtain the following equation:

$$m_1u_1 = m_1v_1 + m_2v_2 \Rightarrow m_1u_1 = m_1v_1 + m_2v_2 \tag{2.3}$$

Solving Eq. 2.1 and 2.2 gives the values of v_2 and v_3 as follows:

$$v_1 = \frac{(m_1 - m_2)u_1}{m_1 + m_2} \tag{2.4}$$

$$v_2 = \frac{2m_1u_1}{m_1 + m_2} \tag{2.5}$$

Equation 2.4 and 2.5 quantify the functional mapping (causal function) between the velocity of the cause (u_1) and the velocity of the effect (v_1 and v_2) and this relaionship can be elegantly expressed in matrix form as follows:

$$\begin{Bmatrix} v_1 \\ v_2 \end{Bmatrix} = \begin{bmatrix} \dfrac{(m_1 - m_2)}{m_1 + m_2} & 0 \\ \dfrac{2m_1}{m_1 + m_2} & 0 \end{bmatrix} \begin{Bmatrix} u_1 \\ u_2 \end{Bmatrix} \tag{2.6}$$

Now that we have described what a causal function is, the next step is to describe what a causal machine is and its relations to rational decision making.

2.5   Causal Machine for Rational Decision Making

A causal machine which was described in Fig. 2.1 is a combination of a causal function and a utility function. Basically a causal function gives an effect for a given cause and each effect has a utility associated with it. Here utility is the value that is derived from an object. A rational decision making process in this context will be a process of identifying the appropriate cause that will maximize the balance of utility. In this chapter we apply rough sets to build a causal function and from this a causal machine and apply this to interstate conflict modeling.

2.6   Interstate Conflict

In this book we illustrate the causal function and machine using the concept of modelling interstate conflict using rough sets technique. The concept of modelling interstate concept is best illustrated in Fig. 2.3.

In Fig. 2.3, there are a number of variables that are used to predict interstate conflict and these include *Allies*, which is a binary measure coded 1 if the members of a dyad are linked by any form of military alliance and 0 if they are not. Another variable is the *contingency* which is also binary, and is coded 1 if both states are geographically contiguous meaning they share a border and 0 if they do not share a border and *Distance* is an interval measure of the distance between the two states' capitals. *Major power* is a binary variable coded 1 if either or both states in the dyad are a major power. A dyad here means each variable is manifestation of relationships between two countries. The variable *Democracy* is measured on a scale where 10 is an extreme democracy and − 10 is an extreme autocracy, and we identify the country with the lowest level of democracy as the weakest link, in this chapter the value of the less democratic country in the dyad is used for our analyses. The variable *Dependency* is the sum of the countries import and export with its partner divided by the Gross Domestic Product of the stronger country and this is a continuous variable measuring the level of economic interdependence (dyadic trade as a portion of a state's gross domestic product) of the less economically dependent state in the dyad. *Capability* is the logarithm, to the base 10, of the ratio of the sum of the total population, the number of people in urban areas, industrial energy consumption, iron and steel production, the number of military personnel in active duty as well as military expenditure in dollars in the last 5 years measured on stronger country to a weaker country. The details of this can be found in a seminal book by Marwala and Lagazio (2011).

The next pertinent question is under what conditions does the model in Fig. 2.3 becomes a causal function. The first condition is that the input should have hap-

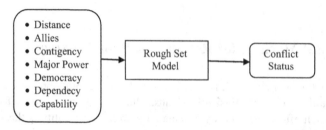

Fig. 2.3 A rough sets based model for conflict prediction

pened or been known before the effect. The second condition is that there has to be a flow of information from the cause to the effect. Taking for an example variable *Distance*, the rational decision maker would have taken into account of the distance from his capital before engaging in militarized conflict. Again with the variable *Allies* the rational decision maker would have taken into account of whether there are any strategic alliances between the countries before engaging in militarized conflict. Likewise, the rational decision maker would have also looked at variables Contingency, Major Power, Dependency, Capability and perhaps Democracy before engaging in militarized conflict. The model in Fig. 2.3 does meet the two criteria for a causal machine.

The data set used in this chapter is the population of politically relevant dyads from 1885 to 2001 which is described in detail and used by Oneal and Russett (2005). The politically relevant population which are contiguous dyads as well as all dyads containing a major power, are selected because they set a hard test for prediction. By neglecting all distant dyads composed of weak states means we ignore much of the influence that variables not very amenable to policy intervention, distance and national power, would exert in the full data set and make the task harder by reducing the predictive power of such variables. Concentrating only on dyads that either include major powers or are contiguous, we test the discriminative power of the predictive models on a difficult situations.

The dependent variable of the models consisted of a binary variable which indicated the onset of a militarized interstate dispute (MID) of any severity after the independent variables have been observed (Maoz 1999). Merely dyads with no dispute or with only the initial year of the militarized conflict, ranging from any severity to war, are included in the analysis.

The next section describes the rough set causal function which is used to model the framework in Fig. 2.3.

2.7   Rough Sets Causal Function

Rough sets technique is a computational intelligence method which is intended to estimate concepts from the observed data. Unlike other computational intelligence methods that are applied to handle uncertainty and imprecision, rough set theory does not require any additional information about the experimental training data such as the statistical probability (Pawlak 1991; Pawlak and Munakata 1996; Crossingham 2007; Crossingham and Marwala 2007; Nelwamondo 2008; Marwala and Lagazio 2011). Rough set theory handles the estimation of sets that are hard to explain with the available information (Ohrn 1999; Ohrn and Rowland 2000; Marwala and Lagazio 2011). It is used mainly to classify uncertainty and uses the upper and lower estimation to handle inconsistent information.

Rough set theory is based on a set of rules, which are described in terms of linguistic variables. Rough sets have been applied to decision analysis, particularly in the analysis of decisions in which there are contradictions (Marwala 2012). Because they are rule-based, rough sets are highly transparent but they are not as accurate. However, they are not good universal estimators and other machine learning techniques such as neural networks are better in their predictions. Thus, in machine learning, there is always a trade-off between prediction accuracy and transparency.

Rough set theory offers a method of reasoning from vague and imprecise data (Goh and Law 2003). The method is based on the assumption that some observed information is in some way associated with some information in the universe of the discourse (Komorowski et al. 1999; Kondo 2006). Objects with the same information are indiscernible in the view of the available information. An elementary set consisting of indiscernible objects forms a basic granule of knowledge. A union of an elementary set is referred to as a *crisp set*, or else, the set is considered to be *rough*. In the next sub-sections, rough set theory is described.

2.7.1 Information System

An information system (Λ), is described as a pair (U, A) where U is a finite set of objects known as the universe and A is a non-empty finite set of attributes as described as follows (Crossingham 2007; Nelwamondo 2008; Marwala 2012; Marwala and Lagazio 2011).

$$\Lambda = (U, A) \tag{2.7}$$

All attributes $a \in A$ have values, which are elements of a set V_a of the attributes a (Dubois 1990; Crossingham 2007; Marwala and Lagazio 2011):

$$a : U \rightarrow V_a \tag{2.8}$$

A rough set is described with a set of attributes and the indiscernibility relation between them. Indiscernibility is explained in the next subsection.

2.7.2 The Indiscernibility Relation

The *indiscernibility relation* is one of the central ideas of rough set theory (Grzymala-Busse and Siddhaye 2004; Pawlak and Skowron 2007; Marwala and Lagazio 2011; Marwala 2012). *Indiscernibility* basically suggests similarity (Goh and Law 2003) and, consequently, these sets of objects are indistinguishable. Given an information system Λ and subset $B \subseteq A$, B the indiscernibility defines a binary relation $I(B)$ on U such that (Pawlak et al. 1988; Ohrn 1999; Ohrn and Rowland 2000; Nelwamondo 2008; Marwala and Lagazio 2011):

$$(x, y) \in I(B)$$
$$\textit{if and only if} \tag{2.9}$$
$$a(x) = a(y)$$

for all $a \in A$ where $a(x)$ symbolizes the value of attribute a for element x. Supposing that U has a finite set of N objects $\{x_1, x_2,..,x_N\}$. Let Q be a finite set of n attributes $\{q_1, q_2,...,q_n\}$ in the same information system Λ, then (Inuiguchi and Miyajima 2007; Crossingham 2007; Nelwamondo 2008; Marwala and Lagazio 2011):

$$\Lambda = \langle U, Q, V, f \rangle \tag{2.10}$$

Here f is the total decision function, known as the *information function*. From the explanation of the indiscernibility relation, two entities have a similarity relation to attribute a if they universally have the same attribute values.

2.7.3   Information Tables and Data Representation

An information table is applied in rough sets theory as a technique for signifying the data. Data in the information table are organized, centered on their condition attributes and decision attributes (D). *Condition attributes* and *decision attributes* are similar to the independent variables and dependent variable (Goh and Law 2003). These attributes are divided into $C \cup D = Q$ and $C \cup D = 0$. Data set is indicated in the table and each object is characterized in an *Information System* (Komorowski et al. 1999).

2.7.4   Decision Rules Induction

Rough sets theory also requires producing decision rules for a given information table. The rules are generally based on condition attributes values (Slezak and Ziarko 2005). The rules are presented in an '*if CONDITION(S)-then DECISION*' format.

2.7.5   The Lower and Upper Approximation of Sets

The lower and upper approximations of sets are defined on the basis of the indiscernibility relation. The *lower approximation* is defined as the collection of cases whose equivalent classes are confined in the cases that need to be estimated, while the *upper approximation* is defined as the collection of classes that are incompletely contained in the set that need to be estimated (Marwala 2012). If the concept X is defined as a set of all cases defined by a specific value of the decision and that any finite union of elementary set, associated with B called a B-definable set (Grzymala-Busse and Siddhaye 2004) then set X can be estimated by two B--definable sets, known as the B-lower estimation denoted by BX and B-upper approximation $\overline{B}X$. The B-lower approximation is defined as (Bazan et al. 2004; Crossingham 2007; Nelwamondo 2008; Marwala 2012):

$$\underline{B}X = \left\{x \in U \,\middle|\, [x]_B \subseteq X\right\} \qquad\qquad (2.11)$$

and the B-upper approximation is defined as (Crossingham 2007; Nelwamondo 2008; Marwala and Lagazio 2011):

$$\overline{B}X = \left\{x \in U \,\middle|\, [x]_B \cap X \neq 0\right\} \qquad\qquad (2.12)$$

There are other approaches that have been described for defining the lower and upper approximations for a completely specified decision table. Some of the popular ones include approximating the lower and upper approximation of X using Eq. 2.11 and 2.12, as follows (Grzymala-Busse 2004; Crossingham 2007; Nelwamondo 2008; Marwala 2012):

$$\cup\left\{[x]_B \,\middle|\, x \in U, [x]_B \subseteq X\right\} \qquad\qquad (2.13)$$

$$\cup\left\{[x]_B \,\middle|\, x \in U, [x]_B \cap X \neq 0\right\} \qquad\qquad (2.14)$$

The definition of definability is revised in situations of incompletely specified tables. In this case, any finite union of characteristic sets of B is called a B-definable set. Three different definitions of approximations have been discussed by Grzymala-Busse and Siddhaye (2004). By letting B be a subset of A of all attributes and $R(B)$ be the characteristic relation of the incomplete decision table with characteristic sets $K(x)$, where $x \in U$, the following can be defined (Grzymala-Busse 2004; Grzymala-Busse 2004; Crossingham 2007; Nelwamondo 2008; Marwala 2012):

$$\underline{B}X = \{x \in U \,|\, K_B(x) \subseteq X\} \qquad\qquad (2.15)$$

and

$$\overline{B}X = \left\{x \in U \,\middle|\, K_B(x) \cap X \neq 0\right\} \qquad\qquad (2.16)$$

Equation 2.15 and 2.16 are known as *singletons*. The subset lower and upper approximations of incompletely specified data sets can then be mathematically defined as (Nelwamondo 2008; Marwala 2012):

$$\cup\left\{K_B(x) \,\middle|\, x \in U, K_B(x) \subseteq X\right\} \qquad\qquad (2.17)$$

and

$$\cup\left\{K_B(x) \,\middle|\, x \in U, K_B(x) \cap X = 0\right\} \qquad\qquad (2.18)$$

Additional information on these approaches can be found in (Grzymala-Busse and Hu 2001; Grzymala-Busse and Siddhaye 2004; Crossingham 2007; Marwala and Lagazio 2011). It can be deduced from these properties that a crisp set is only defined if $\underline{B}(X) = \overline{B}(X)$. *Roughness* is consequently defined as the difference between the upper and the lower approximation.

2.7.6 Set Approximation

A number of properties of rough sets have been presented in the work of Pawlak (1991). An important property of rough set theory is the definability of a rough set (Quafafou 2000). This was explained for the situation when the lower and upper approximations are equal. If this is not the situation, then the target set is un-definable. Some of the distinctive cases of definability are (Pawlak et al. 1988; Crossingham 2007; Nelwamondo 2008; Marwala 2012):

- *Internally definable* set: Here, $\underline{B}X \neq 0$ and $\overline{B}X = U$. The attribute set B has objects that certainly are elements of the target set X, even though there are no objects that can definitively be excluded from the set X.
- *Externally definable* set: Here, $\underline{B}X = 0$ and $\overline{B}X \neq U$. The attribute set B has no objects that certainly are elements of the target set X, even though there are objects that can definitively be excluded from the set X.
- *Totally un-definable* set: Here, $\underline{B}X = 0$ and $\overline{B}X = U$. The attribute set B has no objects that certainly are elements of the target set X, even though there are no objects that can definitively be excluded from the set X.

2.7.7 The Reduct

An additional property of rough sets is the *reduct* which is a concept that defines whether there are attributes B in the information system that are more significant to the knowledge represented in the equivalence class structure than other attributes. It is vital to identify whether there is a subset of attributes which could be completely described by the knowledge in the database. This attribute set is known as the *reduct*.

Shan and Ziarko (1995) formally defined a *reduct* as a subset of attributes $RED \subseteq B$ such that:

- $x_{RED} = x_B$. That is, the equivalence classes that were induced by reducing the attribute set RED are equal to the similar class structure that was induced by the full attribute set B.
- Attribute set RED is minimal because $x_{(RED-A)} \neq x_B$ for any attribute $A \in RED$. Simply, there is no attribute that can be taken away from the set RED without changing the equivalent classes x_B.

Therefore, a reduct can be visualized as a suitable set of features that can adequately express the category's structure. One property of a reduct in an information system is that it is not unique since there may be other subsets of attributes which may still preserve the equivalence class structure conveyed in the information system. The set of characteristics that are common in all reducts is called a *core*.

2.7.8   Boundary Region

The *boundary region*, which can be expressed as the difference $\overline{B}X - \underline{B}X$, is a region which is composed of objects that cannot be included nor excluded as elements of the target set X. Simply, the lower approximation of a target set is an estimation which consists only of those objects which can be positively identified as elements of the set. The upper approximation is a rough approximation and includes objects that may be elements of the target set. The boundary region is the area between the upper and lower approximation.

2.7.9   Rough Membership Functions

A *rough membership function* is a function $\mu_A^x \colon U \rightarrow [0,1]$ that, when applied to object x, quantifies the degree of overlap between set X and the indiscernibility set to which x belongs. The rough membership function is applied to estimate the plausibility and can be defined as (Pawlak 1991; Crossingham 2007; Nelwamondo 2008; Marwala and Lagazio 2011):

$$\mu_A^x(X) = \frac{\left|[x]_B \cap X\right|}{\left|[x]_B\right|} \tag{2.19}$$

The rough membership function can be understood as a fuzzification within the context of rough set approximation. It *confirms* the translation from rough set approximation into membership function. The important aspect of a rough membership function is that it is derived from data (Crossingham 2007).

2.7.10   Discretization Using Equal-Width-Bin (EWB) Partitioning

The methods which allow continuous data to be processed involve discretization. There are several methods available to perform discretization and these include Equal-Width-Bin (EWB) and Equal-Frequency-Bin (EFB) partitioning. Crossingham (2007) applied particle swarm optimization to optimize the discretization process and this is applied in this section.

EWB partitioning divides the range of observed values of an attribute into k equally sized bins (Crossingham 2007). For this chapter, k was taken as four. One notable problem of this method is that it is vulnerable to outliers that may drastically skew the data range. This problem was eliminated through a pre-processing step involving cleaning of the data. The manner in which data can be discretized using EWB is as follows (Grzymala-Busse 2004; Crossingham 2007; Marwala and Lagazio 2011):

- Evaluate the smallest and largest values for each attribute and label these values S and L.
- Write the width of each interval, W, as:

$$W = \frac{L-S}{4} \tag{2.20}$$

- The interval boundaries can be determined as: $S + W$, $S + 2\ W$, $S + 3\ W$. These boundaries can be determined for any number of intervals k, up to the term $S + (k-1)W$.

2.7.11 Rough Set Formulation

The process of modeling the rough set can be classified into these five stages (Crossingham 2007):

1. The first stage is to select the data.
2. The second stage involves pre-processing the data to ensure that it is ready for analysis. This stage involves discretizing the data and removing unnecessary data (cleaning the data).
3. If reducts are considered, the third stage is to use the cleaned data to generate reducts. A *reduct* is the most concise way in which we can discern object classes. In other words, a reduct is the minimal subset of attributes that enables the same classification of elements of the universe as the whole set of attributes. To cope with inconsistencies, lower and upper approximations of decision classes are defined in this stage.
4. Stage four is where the rules are extracted or generated. The rules are usually determined based on condition attribute values. Once the rules are extracted, they can be presented in an '*if* CONDITION(S)-*then* DECISION' format.
5. The fifth and final stage involves testing the newly created rules on a test set. The accuracy must be noted and sent back into the optimization method used in Step 2 and the process will continue until the optimum or highest accuracy is achieved.

The procedure for computing rough sets and extracting rules is given in Algorithm 1 (Crossingham 2007). Once the rules are extracted, they can be tested using a set of testing data. The classification output is expressed as a decision value which lies between 0 and 1. The accuracy of the rough set is determined using the Area Under the receiver operating characteristic Curve (AUC).

2.7.12 Interstate Conflict

In this chapter rough sets were implemented with militarized interstate dispute (MID) data. The COW data are used to generate training and testing sets. The training data set consists of 500 conflicts and 500 non-conflict cases, and the test data consists of 392 conflict data and 392 peace data. We use a balanced training set, with a randomly selected equal number of conflicts and non-conflicts cases, to produce robust classifications and stronger insights on the reasons of conflicts. The data are normalized to fall between 0 and 1. Data were discretized using equal width discretization method. The MID had seven inputs and each was discretized into four bins. Rough sets was used to build a causal function between the 7 MID variables and the conflict status. The results of these implementation decisions for the MID data were 272 True Conflict, 120 False Peace, 291 True Peace and 99 False Conflicts. The detection rate of conflict was 69% whereas the detection rate for peace was 75%.

2.8 Rough Set Causal Machine

Figure 2.3 indicates that the causal machine is when a causal function is used for decision making. Suppose that this causal function is used to make a decision as to whether to invest in Zimbabwe or not. Then the MID between Zimbabwe and country n are calculated and used to estimate the conflict status here called C_n. Then a cumulative conflict is calculated and if C_n is lower than a certain value then one invests in Zimbabwe and if it is not then one does not invest in Zimbabwe. This decision making process is deemed rational because it is based on relevant information and uses logic (rough set model).

Conclusions

This chapter introduced rough set to build a causal function. This causal function was tested on modeling militarized interstate conflict. The rough set was discretized using equal-width bin partitioning. The results obtained showed the detection rate of conflict of 69% as well as the detection rate for peace of 75%.

References

Annus AM, Smith GT, Masters K (2008) Manipulation of thinness and restricting expectancies: further evidence for a causal role of thinness and restricting expectancies in the etiology of eating disorders. Psychol Addict Behav 22(2):278–287

Anonymous (2012) River crossing puzzle. http://www.smart-kit.com/s888/river-crossing-puzzle-hard/. Accessed 28 Feb 2013

Baedke J (2012) Causal explanation beyond the gene: manipulation and causality in epigenetics. Theoria (Spain) 27(2):153–174

Bazan J, Nguyen HS, Szczuka M (2004) A view on rough set concept approximations. Fundam Inform 59:107–118

Bellman RE (1954) Dynamic programming and a new formalism in the calculus of variations. Proc Nat Acad Sci USA 40:231–235

Bertsekas DP (2005) Dynamic programming and optimal control. Athena Scientific, Belmont

Choudhari A, Nandi GC, Choudhari RD (2005) NRC: a neuro-rough classifier for landmine detection. Proceedings of INDICON 2005: an international conference of IEEE India Council, 2005, art. no. 1590121, pp 46–51

Collins J, Hall E, Paul L (2004) Causation and counterfactuals. MIT, Cambridge

Crossingham B (2007) Rough set partitioning using computational intelligence approach. MSc Thesis, University of the Witwatersrand, Johannesburg

Crossingham B, Marwala T (2007) Using optimisation techniques to granulise rough set partitions. Comput Model Life Sci 952:248–257

Czyzewski A, Królikowski R (2001) Neuro-rough control of masking thresholds for audio signal enhancement. Neurocomputing 36:5–27

Dowe TP, (1992) Wesley Salmon's process theory of causality and conserved quantity theory. Philos Sci 59:195–216

Dubois D (1990) Rough fuzzy sets and fuzzy rough sets. Intl J Gen Syst 17:191–209

Ehring D (1986) The transference theory of causation. Synthese 67:249–258

Ferguson J (2004) Brief survey of the history of the calculus of variations and its applications (arXiv:arXiv:math/0402357), Origin: ARXIV

Gelfand IM, Fomin SV (2000) Calculus of variations. In: Silverman RA (ed) (Unabridged repr ed). Dover Publications, Mineola, p 3 (ISBN 978-0486414485), New York

Goh C, Law R (2003) Incorporating the rough sets theory into travel demand analysis. Tour Manage 24:511–517

Grzymala-Busse JW (2004) Three approaches to missing attribute values—a rough set perspective, In: Proceedings of the IEEE 4th intl conf on data mining, pp 57–64

Grzymala-Busse JW, Hu M (2001) A comparison of several approaches to missing attribute values in data mining. Lecture Notes Art Intell 205:378–385

Grzymala-Busse JW, Siddhaye S (2004) Rough set approaches to rule induction from incomplete data. In: Proceedings of the 10th intl conf on info process and manage of uncertainty in knowledge-based syst 2:923–930

Hausman D, Woodward J (2004) Manipulation and the causal markov condition. Philos Sci 71(5):846–856

Hubbard TL (2012) Causal representation and shamanic experience. J Conscious Stud 19(5–6):202–228

Inuiguchi M, Miyajima T (2007) Rough set based rule induction from two decision tables. Eur J Oper Res 181:1540–1553

Kim J (2013) Many ways of qualitative contrast in probabilistic theories of causality. Qual Quant 47(2):1225–1236

Kistler M (1998) Reducing causality to transmission. Erkenntnis 48:1–24

Komorowski J, Pawlak Z, Polkowski L, Skowron A (1999) A rough set perspective on data and knowledge. In: Klösgen W, Zytkow JM, Klosgen W, Zyt J (eds) The handbook of data mining and knowledge discovery. Oxford University, NY

Kondo M (2006) On the structure of generalized rough sets. Info Sci 176:589–600

Liu J-Q, Yamanishi T, Nishimura H, Umehara H, Nobukawa S (2012) On the filtering mechanism of spontaneous signaling causality of brain's default mode network, 6th international conference on soft computing and intelligent systems, and 13th international symposium on advanced intelligence systems, SCIS/ISIS 2012, art. no. 6505127, pp 43–48

Loewer B (2007) Counterfactuals and the second law, In: Price H, Corry R (eds) Causation, physics and the constitution of reality. Oxford: Oxford University Press, pp 293–326

Maoz Z (1999) Dyadic Militarized Interstate Disputes (DYMID1.1) Dataset-Version 1.1. ftp:// spirit.tau.ac.il./zeevmaoz/dyadmid60.xls (Password protected). Accessed Aug 2000

Marwala T (2012) Condition monitoring using computational intelligence methods. Springer, Heidelberg

Marwala T (2014) Causality, correlation and artificial intelligence for rational decision making. World Scientific Publications, Singapore

Marwala T, Crossingham B (2008) Neuro-rough models for modelling HIV. Conference proceedings—IEEE international conference on systems, man and cybernetics, art. no. 4811770, pp 3089–3095

Marwala T, Crossingham B (2009) Bayesian neuro-rough model. ICIC Express Lett 3(2):115–120

Marwala T, Lagazio M (2011) Militarized conflict modeling using computational intelligence techniques London. Springer, UK

Maudlin T (2004) Causation, Counterfactuals, and the third factor. In: Collins J, Hall N, Paul LA (eds) Causation and counterfactuals, Cambridge, MA: MIT Press. Reprinted in The Metaphysics in Physics, 2007. Oxford: Oxford University Press, pp 419–443

Mitroff II, Silvers A (2013) Probabilistic causality. Technol Forecast Soc Change 80(8):1629–1634

Nelwamondo FV (2008) Computational intelligence techniques for missing data imputation. PhD Thesis, University of the Witwatersrand, Johannesburg

Ohrn A (1999) Discernibility and rough sets in medicine: tools and applications. Unpublished PhD Thesis, Norwegian University of Science and Technology

Ohrn A, Rowland T (2000) Rough sets: a knowledge discovery technique for multifactorial medical outcomes. Amer J Phys Med Rehabil 79:100–108

Oneal J, Russett B (2005) Rule of three, let it be? When more really is better. Confl Manage Peace Sci 22:293–210

Park ER, Streck JM, Gareen IF, Ostroff JS, Hyland KA, Rigotti NA, Pajolek H, Nichter M (2014) A qualitative study of lung cancer risk perceptions and smoking beliefs among national lung screening trial participants. Nicotine Tob Res 16(2):166–173

Pawlak Z (1991) Rough sets—theoretical aspects of reasoning about data. Kluwer Academic, Dordrecht

Pawlak Z, Munakata T (1996) Rough control application of rough set theory to control. In: Proceedings of the 4th Euro congr on intelliTechniq and soft comput:209–218

Pawlak Z, Skowron A (2007) Rough sets and boolean reasoning. Info Sci 177:41–73

Pawlak Z, Wong SKM, Ziarko W (1988) Rough sets: probabilistic versus deterministic approach. Int J Man-Mach Stud 29:81–95

Pearl J (2000) Causality: models, reasoning, and inference. Cambridge University Press, Cambridge

Quafafou M (2000) α-RST: a generalization of rough set theory. Info Sci 124:301–316

Ramsahai RR (2013) Probabilistic causality and detecting collections of interdependence patterns. J Royal Stat Soc. Series B: Stat Methodol 75(4):705–723

Rebane G, Pearl J (1987) The recovery of causal poly-trees from statistical data, Proceedings, 3rd workshop on uncertainty in AI, (Seattle, WA), pp 222–228

Sabzevari R, Montazer GHA (2008) An intelligent data mining approach using neuro-rough hybridization to discover hidden knowledge from information systems. J Info Sci Eng 24(4):1111–1126

Salmon W (1984) Scientific explanation and the causal structure of the world. Princeton University, Princeton

Salmon WC (1998) Causality and explanation. Oxford University Press, Oxford

Shamsuddin SM, Jaaman SH, Darus M (2009) Neuro-rough trading rules for mining Kuala Lumpur composite index. Eur J Sci Res 28(2):278–286

Shan N, Ziarko W (1995) Discovering attribute relationships, dependencies and rules by using rough sets. In the proceedings of the 28th hawaii international conference on systems sciences, vol. 3, pp. 293–299

Simiński K (2012) Neuro-rough-fuzzy approach for regression modelling from missing data. Int J Appl Math Computer Sci 22(2):461–476

Slezak D, Ziarko W (2005) The investigation of the Bayesian rough set model. Int J Approx Reason 40:81–91

Sokolovski D, Akhmatskaya E (2012) Causality, 'superluminality', and reshaping in undersized waveguides, ECCOMAS 2012– European congress on computational methods in applied sciences and engineering, e-Book Full Papers, pp 93–103

van Brunt B (2004) The calculus of variations. Springer, Berlin (ISBN 0-387-40247-0)

Verma T, Pearl J (1990) Equivalence and synthesis of causal models, Proceedings of the sixth conference on uncertainty in artificial intelligence, (July, Cambridge, MA), pp 220–227, 1990. (Reprinted In: Bonissone P, Henrion M, Kanal LN and Lemmer JF (Eds) Uncertainty in artificial intelligence 6, Amsterdam: Elsevier Science Publishers, B.V., pp 225–268, 1991)

Wright S (1921) Correlation and causation. J Agric Res 7(3):557–585

Yan F, Xu J-F, Liu XF, Li X-H (2014) Interaction between smoking and CYP2C19*3 polymorphism increased risk of lung cancer in a Chinese population. Tumor Biol 35:1–4 (Article in press)

Chapter 3
Correlation Function for Rational Decision Making: Application to Epileptic Activity

3.1 Introduction

Rational decision making is process of making decisions based on relevant information, logic and in an optimized fashion by maximizing utility. Utility here is defined as the value which is derived from using a good or service minus the cost associated with its expense. One way of making rational decision is to use the concept of correlation to make a rational decision. For example, suppose one knows that if it rains a particular bridge gets flooded and, therefore, one needs to use a different route. Then one can gauge the level of the rain and correlate this to how flooded the river might be and use this information to identify the optimal route that will maximize the balance of utility. Here we use the balance of utility because utility can be both positive and negative. Suppose we compare two choices of moving from A to B. The first choice is to use a car and the second choice is to use a bicycle. The rational way is not just looking at the choice which minimizes time but use also the cost of the action.

In this chapter we use the correlation function to relate variables and use the dependant variable to make rational decisions. In this chapter, a correlation function is defined as a function where the probability of inferering the output given the input where there is no flow of information from the input to the output is greater than zero. In this chapter, we apply support vector machine to build a correlation function that relates information from electroencephalogram (EEG) to the epileptic activity of a patient.

This chapter starts by defining correlation then uses this definition to construct a correlation function. Thereafter, the concept of using the EEG signal of a patient to diagnose epileptic activity is described including relevant features that are used in this regard. Then support vector machine (SVM) which is used to build a correlation function is described and then an experimental investigation on the SVM is conducted in order to test the correlation function.

© Springer International Publishing Switzerland 2014

39

T. Marwala, *Artificial Intelligence Techniques for Rational Decision Making*,
Advanced Information and Knowledge Processing,
DOI 10.1007/978-3-319-11424-8_3

3.2 Defining Correlation

The first step towards the building a correlation machine is to define the concept of correlation. According to *google* dictionary, correlation is defined as "a mutual relationship or connection between two or more things" For example, drinking and driving are correlated to more accidents in the road. In this instance, drinking and driving do not cause accidents but are correlated to accidents. In this book it is proposed that the concept of correlation is one of the elements that can be used for rational decision making. As an example, suppose there is an observation that whenever there are dark clouds it usually rains then one can draw a conclusion that there is a relationship between dark clouds and rain. Based on this observed correlation one can then rationally decide that whenever there are dark clouds then one takes an umbrella when they take a walk to a shop. Taking a walk would be a rational choice if it maximizes the balance of utility when compared to say using a car, or a scooter or a bicycle or crawling to the shop.

The next pertinent question to ask is, why would variables be correlated? This section gives three reasons why variables can be correlated:

- Because there is a causal relationship in the variables.
- When they describe a phenomena with similar characteristics. The spread of epidemics is correlated to the population growth.
- When they have a common ancestor. The genes of siblings are correlated because they have common parents.

The next section describes a correlation function which is a mathematical relationship that relates two or more variables which are only correlated but are not causal.

3.3 Correlation Function

Suppose that we know that variable x is correlated to variable y, then it is possible to create a function that maps the relationship between x and y and this relationship is called a correlation function. In Chap. 2, we defined a causal function as a function that takes an input vector (x) and propagate it into the effect (y) where y happens after x and that there is a flow of information between x and y. This, therefore, implies that if one can still map the relationship between x and y (i.e. that the probability of inferring y from x is greater than zero) but there is no flow of information from x to y, then the function is a correlation function. This correlation function can be appropriately represented mathematically as follows (Marwala 2014):

$$y = f(x) \tag{3.1}$$

Here f is the functional mapping. Even though this equation strictly implies that y is directly obtained from x this is a correlation relationship because either or both of

Fig. 3.1 A relationship
between a causal and correlation function

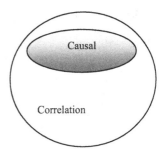

the conditions (1) y happens after x, and (2) there is a flow of information between
x and y are violated. An illustration of a causal function and its relationship to a correlation function is shown in Fig. 3.1.

In this figure a causal function necessarily exhibit correlation relationship
whereas not all correlated relationships are necessarily causal. This is an important
distinction, since it is because of this notion that causality is a special kind of correlation and not the other way round that has caused many great thinkers to confuse
causality with correlation often to the detriment of society.

3.4 Correlation Function for Rational Decision Making

When a correlation function is combined with the concept of utility function, then
a rational decision making framework is born and because it is based on correlation
we will use the term a correlation machine. Figure 3.2 shows a rational decision
framework which is based on a correlation function. In this figure, the first step is to
identify correlation relationships amongst variables. The second step is to identify
the correlation function that describes the variables. Then we relate the variable to
the utility functions representing different choices. Then we choose the cause of
action that maximizes the balance of utility. The idea of making choices based on
the well-known notion of maximizing the balance of utility is based on the theory
of rational choice which assumes that all actions are identifiable, that the outcome
of actions are comparable and that all choices are independent. The next section
describes epileptic activity which is a concept that is modeled in this chapter using
a correlation machine which is based on support vector machine.

3.5 Modelling Epileptic Activity

Epilepsy is a neurological disorder that is principally diagnosed and monitored
using the electroencephalogram (EEG) (Mohamed et al. 2006; Magiorkinis et al.
2010). The EEG is a recording of the variations in the weak electric potentials along

Fig. 3.2 The correlation
machine for rational decision
making

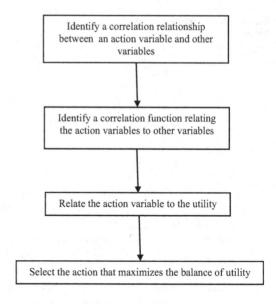

the scalp (Niedermeyer and da Silva 2004). Epileptiform activity is observed in
the EEG as transient waveforms in the form of spikes and sharp waves (SSWs).
Notwithstanding developments in the EEG technology, analysis of EEG records is
typically executed manually by highly trained professionals who visually observe
the EEG for the incidence of spike waveforms. This is a strenuous chore because
the EEG with its numerous recording channels can yield enormous quantity of data.
Automating the epileptic activity detection process with a computer-based pattern
recognition system yields considerable time saving.

Due to the complexity of interpreting EEG signal because of issues such as inter-
subject variability of patients caused by age, level of consciousness and experimental conditions, it has become important to use machines to interpret the EEG signals.
In order to do this it is important to understand the relationship between the EEG
signals and the status of epileptic activity of a patient. In this chapter it is proposed
that this relationships can be accurately modelled using the SVM correlation function. The relationship between EEG signal and epileptic activity is a correlation
relationship because even though there might be an argument that epileptic activity
causes the EEG signal to be of a certain type, the EEG signal does not precede the
epileptic activity and thus the relationship between the EEG signal and epileptic activity where the EEG signal is an independent variable while the EEG signal is the
independent variable is not a causal relationship and by the process of elimination
is thus a correlation function.

This correlation relationship between the EEG signal and epileptic activity has
been successfully modelled using techniques such as neural networks (Gabor and
Seyal 1992; Jando et al. 1993; Kalayci and Ozdamar 1995; Mohamed 2003; Mohamed et al. 2006) and support vector machine as it is done in this chapter (Fernandes 2011).

There are a number of abnormal EEG activities associated with epilepsy and these are during seizures or between seizures (Gotman et al. 1978) and in this chapter as it was done by Mohamed (2003), only during seizures epileptic activity is studied because it is frequently used for clinical diagnosis of epilepsy. The types of waveforms of epileptic seizure are (Mohamed 2003):

1. *Spikes and Sharp Waves* (*SSWs*): These are spasmodic, transient waveforms and have high amplitudes compared to background activity and sharp peaks. Polyspikes is the term given to multiple repetitive spikes that are repeated at a frequency of approximately 20 Hz.
2. *Spike and Wave Complex*: SSWs are occasionally tailed by a slow wave of frequency of less than 4 Hz and many artifacts are also often present in the EEG that is not generated by cerebral activity such as the movement of eye and electrical interference. The diagnosis system should distinguish between SSWs and artifacts. In building a correlation function, four different types of features have been extracted from the EEG and these are (Mohamed 2003):

   a. *Fast Fourier Transform* (*FFT*): The real and imaginary parts of the FFT of an EEG data segment can be used as inputs to a correlation function and they express a spectral representation of an EEG data segment. Fourier transform is a process which is based on the fact that a periodic function can be estimated by a series of cosine and sine functions (Fourier 1878; Katznelson 1976; Gonzalez-Velasco 1992; Boyce and DiPrima 2005) and this series is known as Fourier series. From the Fourier series one can be able to transform a signal in the time domain into the frequency domain and vice versa. Time domain data can thus be transformed into the frequency domain as follows:

$$X(w) = \frac{1}{2\pi} \int_{-\infty}^{\infty} x(t)e^{-i\omega t} dt \qquad (3.2)$$

Here $x(t)$ is the signal x as function of time t, $X(\omega)$ is the signal X in the frequency domain as a function of frequency ω and i is equal to $\sqrt{-1}$. The Fast Fourier transform is a computationally efficient technique for transforming time domain data into the frequency domain and more details on this are Appendix A.

   b. *Autoregressive Modeling*: The Autoregressive Model has been applied to identify features from the EEG data in a number of EEG classification systems by fitting an EEG data segment to a parametric model (Sharma and Roy 1997; Pardey et al. 1996).
   c. *Wavelet Transform (WT)*: The EEG is a highly non-stationary signal and therefore autoregressive modeling and FFT are not ideal. The WT is appropriate for non-stationary signals and executes a multi-resolution analysis of a signal to give features with temporal and spectral information (Marwala 2002; Tao and Tao 2010). As described by Marwala (2000), the wavelet transform of a signal is an illustration of a timescale decomposition, which highlights the local features of a signal. Wavelets occur in sets of functions that are defined

by dilation, which controls the scaling parameter, and translation, which controls the position of a single function known as the mother wavelet, $w(t)$. In general, each set of wavelet can be written as follows (Newland 1993):

$$W_{ab}(t) = \frac{1}{\sqrt{a}} w\left(\frac{t-b}{a}\right) \tag{3.2}$$

where b = translation parameter, which localizes the wavelet function in the time domain; a = dilation parameter, defining the analyzing window stretching; and w = mother wavelet function. The continuous WT of a signal $x(t)$ is defined as follows (Newland 1993):

$$W(2^j + k) = 2^j \int_{-\infty}^{\infty} x(t)w^*(2^j t - k)dt \tag{3.3}$$

where w^* = complex conjugate of the basic wavelet function; j is called the level (scale) and determines how many wavelets are needed to cover the mother wavelet and is the same as a frequency varying in harmonics; and k determines the position of the wavelet and gives an indication of time. The length of the data in the time domain must be an integer power of two.

The wavelets are organized into a sequence of levels 2^j, where j is from 1 to n -1. Equations 3.2 and 3.3 are valid for $0 \le k$ and $0 \le k \le 2^j - 1$. The WT in this chapter is from the orthogonal wavelet family (Daubechie 1991) and is defined by Newland (1993) as:

$$w(t) = \frac{(e^{i4\pi t} - e^{i2\pi t})}{i2\pi t} \tag{3.4}$$

The WT may also be formulated by transforming the signal $x(t)$ and the wavelet function into the frequency domain as follows (Newland 1993):

$$W(j,k) = \int_{2\pi 2^j}^{4\pi 2^j} X(\omega)e^{i\omega k 2^j} d\omega \tag{3.5}$$

Further details on the WT are Appendix A.

3. *Raw EEG*: The raw EEG signal can be used as input to a correlation function (Kalayci and Özdamar 1998) and this is important because raw EEG data contain all information while extracted features may not represent all the information.

In this chapter, we use five input variables to create a correlation function that maps these variables to epileptic activity and this is shown in Fig. 3.3. The five input variables are: Maximum Laypunov Exponent of Window; Standard Deviation of a Window; Standard Deviation of a Window relative to Chanel; Maximum Coefficient from DWT Analysis; and Maximum Harmonic of Window.

The Lyapunov exponent in the field of dynamical system is a measure that describes the rate of separation of infinitesimally close trajectories (Temam 1988). The next section describes support vector machine which is used to create a correlation function.

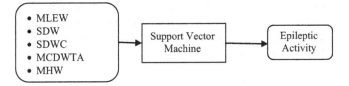

Fig. 3.3 A correlation function based on support vector machines. Key: *MLEW* Maximum Lay-punov Exponent of Window, *SDW* Standard Deviation of a Window, *SDWC* Standard Deviation of a Window relative to Chanel, *MCDWTA* Maximum Coefficient from DWT Analysis, *MHW* Maximum Harmonic of Window

3.6   Support Vector Machine (SVM) Correlation Function

Support vector machines are supervised learning procedures applied primarily for classification and are formulated using statistical learning theory (Vapnik 1998). Shen et al. (2005) used SVMs successfully for color image watermarking whereas Marwala et al. (2006) successfully applied SVMs for the fault classification of mechanical systems and Msiza et al. (2007) successfully applied SVMs in forecasting water demand time series. Chen et al. (2011) successfully estimated monthly solar radiation using SVMs whereas Yeh et al. (2011) successfully applied SVMs to identify counterfeit banknotes. Kumar et al. (2014) successfully applied SVM for epileptic seizure detection using DWT based fuzzy approximate entropy whereas Farquad et al. (2014) successfully applied support vector machine for churn prediction. Huang et al. (2014) successfully applied SVMs for monthly stream flow prediction whereas Sattari et al. (2014) successfully applied support vector machine to determine the speed of sound in ionic. Pires and Marwala (2004) used SVMs for option pricing and further extended this to the Bayesian framework while Gidudu et al. (2007) applied SVMs in image classification.

   In this chapter, SVM is used to construct a correlation function which classifies epileptic activity from EEG data. In this regard a data point is conceptualized as a p-dimensional vector in support vector machines and the objective is to divide such points with a p-1-dimensional hyper-plane and the resulting system is a linear classifier. There are a number of hyper-planes that can be applied to construct a support vector machine. The chosen hyper-plane can be selected to maximize the distance from it to the nearest data point on both sides and it called the *maximum-margin hyper-plane*. This is done by estimating a function $f : R^N \rightarrow \{-1,1\}$ dependent on input-output training data which are created from an independently and identically distributed unknown probability distribution $P(\{x\},y)$ so that f is capable of classifying concealed $(\{x\},y)$ data (Müller et al. 2001; Habtemariam 2006). The mathematical expression of the desired function which minimizes the expected error (risk) can be expressed as follows (Habtemariam 2006; Habtemariam et al. 2005; Marwala and Lagazio 2011):

$$R[f] = \int l(f(\{x\}), y) dP(\{x\}, y) \qquad (3.5)$$

here l is a loss function (Müller et al. 2001). Equation 3.5 cannot be solved directly because the probability distribution P is unknown and, therefore, the estimation is done by identifying an upper bound for the risk function using (Vapnik 1995; Müller et al. 2001; Marwala and Lagazio 2011):

$$R[f] = R[f]_{emp} + \sqrt{\frac{h\left(\ln\frac{2n}{h}+1\right) - \ln\left(\frac{\delta}{4}\right)}{n}} \qquad (3.6)$$

here $h \in N^+$ is the Vapnik-Chervonenkis (*VC*) dimension of $f \in F$ and $\delta > 0$. The *VC* dimension of a function class F is the biggest number of h co-ordinates that can be distributed in all possible ways by means of functions of that class (Vapnik 1995). The empirical error $R[f]_{emp}$ which is a training error is expressed as (Habtemariam 2006; Marwala and Lagazio 2011):

$$R[f]_{emp} = \frac{1}{n}\sum_{i+1}^{n} l(f(x_i), y_i) \qquad (3.7)$$

Supposing that the training example can be linearly separated by a hyper-plane as follows (Habtemariam 2006; Marwala and Lagazio 2011):

$$f(x) = \langle w, \{x\} \rangle + b \quad \text{with} \quad w \in \chi \ , \quad b \in \Re \qquad (3.8)$$

here $\langle .,. \rangle$ represents the dot product, $\{w\}$ represents an adjustable weight vector and $\{b\}$ is an offset. The learning process invented by Vapnik and Lerner (1963) determines the hyper-plane with maximum margin of separation from the class of dividing hyper-planes. Nonetheless, for the reason that practical data normally exhibit complex properties which cannot be separated linearly, more complex classifiers are necessary. The concept of linear classifiers in a feature space is introduced to avoid the complexity of the non-linear classifiers. SVMs identify a linear separating hyper-plane by mapping the input space into a higher dimensional feature space F and by so doing each training example x_i is substituted by $\Phi(x_i)$ to give (Habtemariam 2006; Marwala and Lagazio 2011):

$$Y_i((\{w\}.\Phi(\{x\}_i) + b), i = 1, 2, \ldots, n \qquad (3.9)$$

The *VC* dimension h in the feature space F is constrained subject to $h \leq \| W \|^2 R^2 + 1$ where R is the radius of the smallest sphere around the training data (Müller et al. 2001; Habtemariam 2006; Marwala and Lagazio 2011). Consequently, minimizing the expected risk can be expressed as an optimization problem as follows (Burges 1998; Müller et al. 2001; Schölkopf and Smola 2003; Marwala and Lagazio 2011):

$$\text{Minimize } (\{w\}, b)\frac{1}{2}\| \{w\} \|^2 \qquad (3.10)$$

Subject to:

$$c_i(\{w\},\{x\}_i - b) \geqslant 1, i = 1,\ldots,n \tag{3.11}$$

Equations 3.10 and 3.11 are co-operatively known as the *quadratic programming problem* for the reason that it is the problem of optimizing a quadratic function of a number of variables subject to linear constraints on these variables (Schölkopf and Smola 2003; Marwala and Lagazio 2011). Using the expressions (Marwala and Lagazio 2011):

$$\|\{w\}\|^2 = w.w \tag{3.12}$$

$$\{w\} = \sum_{i=0}^{n} \alpha_i c_i \{x\}_i \tag{3.13}$$

The dual problem of the support vector machines can be expressed as follows by maximizing α_i in Lagrangian form (Schölkopf and Smola 2003; Marwala and Lagazio 2011):

$$\begin{aligned} L(\alpha) &= \sum_{i=1}^{n} \alpha_i - \frac{1}{2} \sum_{i,j} \alpha_i \alpha_j c_i c_j \{x\}_i^T \{x\}_j \\ &= \sum_{i=1}^{n} \alpha_i - \frac{1}{2} \sum_{i,j} \alpha_i \alpha_j c_i c_j k(\{x\}_i,\{x\}_j), i = 1,\ldots,n \end{aligned} \tag{3.14}$$

Subject to:

$$\alpha_i \geq 0, i = 1,\ldots,n \tag{3.15}$$

As well as the constraint from the minimization in b:

$$\alpha_i \geq 0, i = 1,\ldots,n \tag{3.16}$$

subject to the following constraints:

$$\sum_{i=1}^{n} \alpha_i c_i = 0 \tag{3.17}$$

Where the kernel is (Müller et al. 2001):

$$k(\{x\}_i,\{x\}_j) = \{x\}_i \cdot \{x\}_j \tag{3.18}$$

A) Soft margin Cortes and Vapnik (1995) presented an improved maximum margin concept that takes into consideration of the misclassified data points. If there is no hyper-plane that can exactly separate different classes of data points, the *Soft*

Margin technique will select a hyper-plane that divides the data points as efficiently as possible and yet maximizing the distance to the nearest data points. The method uses slack variables, γ_i to estimate misclassification rate of the data point using (Cortes and Vapnik 1995; Marwala and Lagazio 2011):

$$c_i\left(\{w\}\cdot\{x\}_i - b\right) \geq 1 - \gamma_i, 1 \leq i \leq n \tag{3.19}$$

A function that penalizes non-zero γ_i augments of the objective function strike a compromise between a large margin and a small error penalty. Suppose a linear penalty function used then an optimization problem that minimizes $\{w\}$ and γ_i is as follows (Cortes and Vapnik 1995; Marwala and Lagazio 2011):

$$\frac{1}{2}\|\{w\}\|^2 + C\sum_{i=1}^{n}\gamma_i \tag{3.20}$$

subject to:

$$c_i\left(\{w\}\cdot\{x\}_i - b\right) \geq 1 - \gamma_i, \gamma_i \geq 0, i = 1,...,n \tag{3.21}$$

Here C is the capacity. Equations 3.20 and 3.21 can be expressed in the Lagrangian form by optimizing the following equation in terms of $\{w\}, \gamma, b, \alpha$ and β (Cortes and Vapnik 1995; Marwala and Lagazio 2011):

$$\min_{\{w\},\gamma,b}\max_{\alpha,\beta}$$
$$\left\{\frac{1}{2}\|\{w\}\|^2 + C\sum_{i=1}^{n}\gamma_i - \sum_{i=1}^{n}\alpha_i\left[c_i\left(\{w\}\cdot\{x\}_i - b\right) - 1 + \gamma_i\right] - \sum_{i=1}^{n}\beta_i\gamma_i\right\} \tag{3.22}$$

where $\alpha_i, \beta_i \geq 0$.

The objective of a linear penalty function is that the slack variables are eliminated from the dual problem and consequently C only manifest itself as a secondary constraint on the Lagrange multipliers. The application of non-linear penalty function to minimize outliers effect on the classifier causes the optimization problem to be non-convex and, therefore, difficult to solve for a global optimum solution.

B) Non-linear classification The kernel trick is applied in support vector machines to transform a linear SVM procedure to a non-linear SVM classifiers (Aizerman et al. 1964; Boser et al. 1992). In the kernel trick the dot product is substituted by a non-linear kernel function to appropriate the maximum-margin hyper-plane in a transformed feature space. Even though this dot product transformation may well be non-linear, the transformed space is usually in high dimensions. Let's say, when a Gaussian radial basis function kernel is applied, the consequential feature space is a Hilbert space of infinite dimension. Some useful kernel functions include (Vapnik 1995; Müller et al. 2001):

- Radial Basis Function
- Polynomial (homogeneous)
- Polynomial (inhomogeneous)
- Hyperbolic tangent.

The variables of the maximum-margin hyper-plane are estimated by optimizing the objective function using an interior point technique that approximates a solution for the Karush-Kuhn-Tucker (KKT) conditions of the primal and dual problems (Kuhn and Tucker 1951; Karush 1939). To improve the effectiveness of the implemented procedure solving a linear system with large kernel matrix was avoided and a low rank approximate to the matrix was used for the kernel trick. The Karush-Kuhn-Tucker conditions are used to optimize a non-linear programming problem to satisfy a specific regularity condition in the problem:

$$Minimize : f(\{x\}) \tag{3.23}$$

subject to:

$$g_i(\{x\}) \le 0; h_j(\{x\}) = 0 \tag{3.24}$$

Where, g_i is the ith inequality constraint and h_i is the ith equality constraint. The Karush-Kuhn-Tucker procedure is useful for the inequality constraints as it generalizes the method of Lagrange multipliers which is useful for mainly equality constraints. The necessary conditions for the KKT can be expressed as follows (Kuhn and Tucker 1951; Karush 1939):

Stationary:

$$\nabla f(\{x^*\}) + \sum_{i=1}^{m} \mu_i \nabla g_i(\{x^*\}) + \sum_{j=1}^{l} \lambda_j \nabla h_j(\{x^*\}) = 0, i = 1,...,m; j = 1,...,l \tag{3.25}$$

Primal and dual feasibility as well as complementary slackness:

$$g_i(\{x^*\}) \le 0, i = 1,...,m$$
$$h_j(\{x^*\}) = 0; j = 1,...,l \tag{3.26}$$
$$\mu_i \ge 0, i = 1,...,m$$
$$\mu_i g_i(\{x^*\}) = 0, i = 1,...,m$$

The KKT method can be considered to be a generalized Lagrangian method by setting $m = 0$. For certain conditions the necessary conditions are simultaneously also sufficient for optimization. Nevertheless, in many states the necessary conditions are not sufficient for optimization and information like the second derivative is necessary. The necessary conditions are sufficient for optimization if the objective function f and the inequality constraints g_j are continuously differentiable convex functions and the equality constraints g_j are functions which have constant gradients.

3.7   Application to Epileptic Activity

The experimental data that was used to construct a correlation function in this chapter was logged from eleven patients with 12-bit accuracy at a sampling frequency of 200 Hz using a Nihon-Kohden 2100 digital EEG system (Mohamed 2003; Mohamed et al. 2006). In the sample gathered, nine of the patients' EEG had epileptiform activity and the other two presented normal brain activity. As the data is gathered from quite a few different patients, the capacity of detectors to endure inter-patient variations in the EEG can be investigated. Data was designated as exhibiting either epileptic activity, artifact activity or normal background activity by two EEG engineers of extensive experience. Data and their corresponding labels were identified from the dataset, of which half exhibited epileptiform activity and the balance had artifact or background activity. The labels were applied to determine targets for support vector machine training and validation. Three different experiments were investigated with the intention of assessing whether the approaches described in this chapter are beneficial in the framework of epileptic detection using a correlation function. The results indicating the sample data from the EEG data are indicated in Fig. 3.4 (Mohamed 2003; Mohamed et al. 2006).

The summary of the implementation of support vector machines is shown in Fig. 3.5. The EEG data in Fig. 3.4 were input into the support vector machine which used radial basis function kernel function (Chang et al. 2010; Vert et al. 2004). The SVM correlation function was implemented with a soft margin of 0.01 and the results are shown in Table 3.1. Sensitivity which is also called true positive rate is defined in this chapter as the proportion of epileptic activities which are classified correctly and in this example 77 % of epileptic cases where classified correctly (Pewsner et al. 2004).

Specificity which is also called true negative rate is defined in this chapter as the proportion of non-epileptic cases that are classified correctly and in this example this is 77 % (Macmillan and Creelman 2004). These results are similar to the results obtained by Fernandes (2011). The overall classification accuracy defined as the proportion of cases classified correctly was 80 % whereas 4.5 % could not be classified one way or another.

These results demonstrate that it is quite possible for support vector machines to be used to construct a correlation function which defines the diagnostics capability of epileptic activity.

Fig. 3.4 Illustration of epileptic spikes **a**, sharp waves **b** and spike and wave complex of EEG

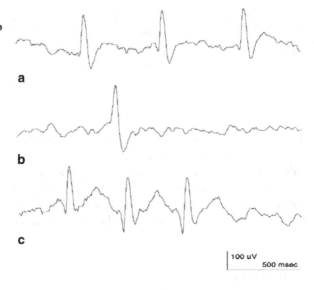

Fig. 3.5 A support vector machine correlation function for mapping the EEG data to epileptic activity

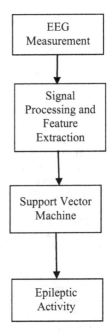

Conclusions

This chapter defined a correlation function which is a function that is able to relate a group of two sets of variables which have no causal relationship but merely a correlation relationship. It was also described that the correlation function is a generalized form of relationship where the causal function is just a special case where

Table 3.1 Results obtained for predicting epileptic activity using support vector machine

Measure (%)	Results (soft margin of 0.1) (%)
Sensitivity	77
Specificity	86
Accuracy	80
Unclassifiable	4.5

there is a flow of information from the cause to the effect. In this chapter, the correlation function is then implemented using support vector machine to create a correlation function which is useful for epileptic detection. This support vector machine correlation function was based on EEG data which was processed using various signal processing methods. The results obtained indicated that it is quite possible to build a causal function which is based on support vector machine.

References

Aizerman M, Braverman E, Rozonoer L (1964) Theoretical foundations of the potential function method in pattern recognition learning. Automat Rem Contr 25:821–837

Boser BE, Guyon IM, Vapnik VN (1992) A training algorithm for optimal margin classifiers. In: Haussler D (ed) 5th annual ACM workshop on COLT. ACM, Pittsburgh

Boyce WE, DiPrima RC (2005) Elementary differential equations and boundary value problems, 8th edn. Wiley, New Jersey

Burges C (1998) A tutorial on support vector machines for pattern recognition. Data Min Knowl Disc 2:121–167

Chang Y-W, Hsieh C-J, Chang K-W, Ringgaard M, Lin C-J (2010) Training and testing low-degree polynomial data mappings via linear SVM. J Mach Learn Res 11:1471–1490

Chen JL, Liu HB, Wu W, Xie DT (2011) Estimation of monthly solar radiation from measured temperatures using support vector machines—a case study. Renew Energy 36:413–420

Cortes C, Vapnik V (1995) Support-vector networks. Mach Learn 20:273–297

Daubechie I (1991) The wavelet transform, time-frequency localization, and signal processing. IEEE Trans Inf Theory 36:961–1005

Farquad MAH, Ravi V, Raju SB (2014) Churn prediction using comprehensible support vector machine: an analytical CRM application. Appl Soft Comput 19:31–40

Fernandes M (2011) SVM to automatically detect epileptic patterns in EEG. Dissertation, University of the Witwatersrand Master

Fourier J (1822) Théorie Analytique de la Chaleur. English edition: Fourier J (2003) The analytical theory of heat (trans: freeman A (1878)) Dover publications. ISBN 0-486-49531-0, unabridged republication of the 1878 English translation by Alexander freeman of fourier's work

Gabor AJ, Seyal M (1992) Automated interictal EEG spike detection using artificial neural networks, elecroenceph. Clin Neurophysiol 83:271–280

Gidudu A, Hulley G, Marwala T (2007) Image classification using SVMs: one-against-one vs one-against-all. Proceeding of the 28th Asian conference on remote sensing:CD-Rom

Gonzalez-Velasco EA (1992) Connections in mathematical analysis: the case of fourier series. Am Math Mon 99(5):427–441. doi:10.2307/2325087

Gotman J, Gloor P, Schaul N (1978) Comparison of traditional reading of the EEG and automatic recognition of interictal epileptic activity, elecroenceph. Clin Neurophysiol 44:48–60

Habtemariam E (2006) Artificial intelligence for conflict management. Master thesis, University of the Witwatersrand, Johannesburg

Habtemariam E, Marwala T, Lagazio M (2005) Artificial intelligence for conflict management. In: Proceedings of the IEEE international joint conference on neural networks, pp 2583–2588

Huang S, Chang J, Huang Q, Chen Y (2014) Monthly streamflow prediction using modified EMD-based support vector machine. J Hydrol 511:764–775

Jando G, Siegel RM, Horvath Z, Buzsaki G (1993) Pattern recognition of the electroencephalogram by artificial neural networks, elecroenceph. Clin Neurophysiol 86:100–109

Kalayci T, Ozdamar O (1995) Wavelet preprocessing for automated neural network detection of EEG spikes. IEEE Eng Med Biol Soc 14:160–166

Kalayci T, Ozdamar O (1998) Detection of spikes with artificial neural networks using raw EEG. Comput Biomed Res 31:122–142

Karush W (1939) Minima of functions of several variables with inequalities as side constraints. MSc thesis, University of Chicago

Katznelson Y (1976) An introduction to harmonic analysis (Second corrected ed.). Dover Publications, Inc, New York

Kuhn HW, Tucker AW (1951) Nonlinear programming. In: Proceeding of 2nd Berkeley Symposium, pp 481–492

Kumar Y, Dewal ML, Anand RS (2014) Epileptic seizure detection using DWT based fuzzy approximate entropy and support vector machine. Neurocomputing 133:271–279

Macmillan NA, Creelman CD (2004) Detection theory: a user's guide. Psychology Press. Florida, USA

Magiorkinis E, Kalliopi S, Diamantis A (2010) Hallmarks in the history of epilepsy: epilepsy in antiquity. Epilepsy Behav 17(1):103–108. doi:10.1016/j.yebeh.2009.10.023

Marwala T (2000) On damage identification using a committee of neural networks. J Eng Mech-ASCE 126:43–50

Marwala T (2002) Finite element updating using wavelet data and genetic algorithm. AIAA J Aircraft 39:709–711

Marwala T (2014) Causality, correlation and artificial intelligence for rational decision making. World Scientific Publications, Singapore

Marwala T, Lagazio M (2011) Militarized conflict modeling using computational intelligence techniques. Springer-Verlag, Heidelberg

Marwala T, Chakraverty S, Mahola U (2006) Fault classification using multi-layer perceptrons and support vector machines. Intl J Eng Simul 7:29–35

Mohamed N (2003) Detection of epileptic activity in the EEG using artificial neural networks. Dissertation, University of the Witwatersrand Master

Mohamed N, Rubin D, Marwala T (2006) Detection of epileptiform activity in human EEG signals using bayesian neural networks. Neural Inf Process Lett Rev 10:1–10

Msiza IS, Nelwamondo FV, Marwala T (2007) Artificial neural networks and support vector machines for water demand time series forecasting. In: Proceedings of the IEEE International conference on systems, man, and cybernetics, pp 638–643

Müller KR, Mika S, Ratsch G, Tsuda K, Scholkopf B (2001) An introduction to kernel-based learning algorithms. IEEE Trans Neural Netw 12:181–201

Newland DE (1993) An introduction to random vibration, spectral and wavelet analysis, 3rd edn. Longman, Harlow, and John Wiley, New York

Niedermeyer E, da Silva FL (2004) Electroencephalography: basic principles, clinical applications, and related fields, Lippincott Williams and Wilkins, Philadelphia

Pardey J, Roberts S, Tarassenko L (1996) A review of parametric modelling techniques for EEG analysis. Med Eng Physics 18:2–11

Pewsner D, Battaglia M, Minder C, Marx A, Bucher HC, Egger M (2004) Ruling a diagnosis in or out with "SpPIn" and "SnNOut": a note of caution. BMJ (Clinical research ed.) 329(7459):209–213. doi:10.1136/bmj.329.7459.209. (PMC 487735. PMID 15271832)

Pires M, Marwala T (2004) Option pricing using neural networks and support vector machines. In: Proceedings of the IEEE International conference on systems, man, and cybernetics, pp 1279–1285

Sattari M, Gharagheizi F, Ilani-Kashkouli P, Mohammadi AH, Ramjugernath D (2014) Determination of the speed of sound in ionic liquids using a least squares support vector machine group contribution method. Fluid Phase Equilibr 367:188–193

Schölkopf B, Smola AJ (2003) A short introduction to learning with kernels. In: Mendelson S, Smola AJ (eds) Proceedings of the machine learning summer school. Springer, Berlin

Sharma A, Roy RJ (1997) Design of a recognition system to predict movement during anaesthesia. IEEE Trans Biomed Eng 44:505–511

Shen R, Fu Y, Lu H (2005) A novel image watermarking scheme based on support vector regression. J Syst Software 78:1–8

Tao X, Tao W (2010) Cutting tool wear identification based on wavelet package and SVM. In: Proceedings of the world congress on intelligent control and automation, pp 5953–5957

Temam R (1988) Infinite dimensional dynamical systems in mechanics and physics. Springer-Verlag, Cambridge

Vapnik V (1995) The nature of statistical learning theory. Springer Verlag, New York

Vapnik V (1998) Statistical learning theory. Wiley-Interscience, New York

Vapnik V, Lerner A (1963) Pattern recognition using generalized portrait method. Automat Rem Contr 24:774–780

Vert J-P, Tsuda K, Schölkopf B (2004) A primer on kernel methods. Kernel methods in computational biology. MIT Press Cambridge Massachusetts, pp 35–70, key: citeulike: 1378324

Yeh CY, Su WP, Lee SJ (2011) Employing multiple-kernel support vector machines for counterfeit banknote recognition. Appl Soft Comput 11:1439–1447

Chapter 4
Missing Data Approaches for Rational Decision Making: Application to Antenatal Data

4.1 Introduction

The issue of missing data estimation framework and the use of this framework for rational decision making is an interesting proposition that requires further investigation. The field of data analysis where datasets are usually incomplete has prompted researchers to come up with many innovative algorithms some of which can be found in Ho et al. (2001), Huang and Zhu (2002), Harel (2007), Marwala (2009), Reiter (2008) and Beunckens et al. (2008).

This section describes methods that have been applied for missing data estimation. In the 1970s, missing data problems were handled by logically inferring missing variables from observed data. A formal technique for estimating missing data was proposed in 1976 and shortly thereafter by Dempster et al. (1977) who proposed the Expectation Maximization (EM) algorithm. Thereafter, Little and Rubin (1987) introduced Multiple Imputations which became possible due to advances in computational capability (Rubin 1987; Faris et al. 2002; Sartori et al. 2005; Donders et al. 2006; Gad and Ahmed 2006).

From the 1990s numerous techniques for approximating missing data were proposed and applied to a diverse areas (Gabrys 2002; Marwala and Chakraverty 2006). Some of the methods that have been proposed for missing data estimation were based on artificial intelligence methods such as neural networks, support vector machines, genetic algorithm and simulated annealing (Dhlamini et al. 2006; Nelwamondo 2008; Junninen et al. 2007; Abdella 2005; Abdella and Marwala 2005, 2006).

There are many techniques that have been proposed to handle missing data many of which are not relevant for the problem at hand i.e. that of rational decision making via missing data approach. These methods that are not relevant include list deletion, pairwise deletion, simple rule prediction, mean substitution, hot deck imputation and cold deck imputation (Marwala 2009). Examples of missing data estimation methods which are relevant for rational decision making include stochastic imputation which is a type of regression imputation that replicates the uncertainty of the

© Springer International Publishing Switzerland 2014
T. Marwala, *Artificial Intelligence Techniques for Rational Decision Making,*
Advanced Information and Knowledge Processing,
DOI 10.1007/978-3-319-11424-8_4

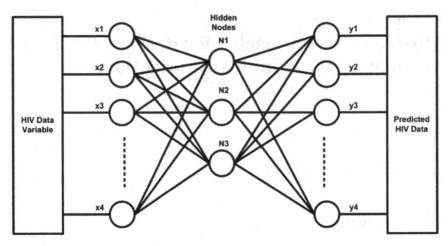

Fig. 4.1. Auto-associative neural network

predicted values and can be viewed as an extension of the simple rule prediction. Computational intelligence based methods can be considered as an example of this class (Nelwamondo 2008) and this method is also known as Multiple Imputation (Rubin 1987) and merges statistical techniques by producing a maximum likelihood based covariance matrix and a vector of means. Multiple Imputation involves drawing missing values from the posterior distribution of the missing values given the observed values and is attained by averaging the posterior distribution for the complete data over the predictive distribution of the missing data. This chapter applies the autoassociative neural network and genetic algorithm for missing data estimation and applies this for rational decision making and these techniques will be described in detail later in the chapter. The next section describes the missing data approach adopted in this chapter.

4.2   Missing Data Approach

The autoassociative neural network missing data estimation method proposed in this chapter for rational decision making involves a model which is created to reproduce its input as the output. As an example, suppose there is a set of data from database and it is a vector of five components. Then an autoassociative model can be created so that every time this vector is presented a corresponding vector identical (nearly) to it is predicted. This model can be any viable mathematical model. Fig. 4.1 shows such a model which is based on multi-layer perceptron neural networks (Marwala 2009). Other models which can be used in this regard include support vector machines and rough sets (Marwala 2009).

This model can be written as follows (Marwala 2009):

$$\{y\} = f(\{x\},\{w\}) \tag{4.1}$$

In Eq. 4.1, $\{y\}$ is the output vector, $\{x\}$ is the input vector and $\{w\}$ is the vector of network weights. For the reason that in this framework a model is created to estimate its own input vector, the input vector $\{x\}$ is approximately equal to output vector $\{y\}$ and therefore $\{x\} \approx \{y\}$. The input vector $\{x\}$ and output vector $\{y\}$ are not always exactly equal and, consequently, an error function can be written as the difference between the input and output vectors as follows:

$$\{e\} = \{x\} - \{y\} \tag{4.2}$$

Substituting the value of $\{y\}$ from Eq. 4.1 into Eq. 4.2 the following equation is obtained:

$$\{e\} = \{x\} - f(\{x\},\{w\}) \tag{4.3}$$

In order for the error to be minimized to be non-negative, the error function in Eq. 4.3 can be rewritten as:

$$\{e\} = (\{x\} - f(\{x\},\{w\}))^2 \tag{4.4}$$

In the case of estimation, some of the values for the input vector $\{x\}$ may not be available. Therefore, the input vector elements can be divided into $\{x\}$ known vector represented by $\{x_k\}$ and $\{x\}$ unknown vector represented by $\{x_u\}$. Rearranging Eq. 4.4 in terms of $\{x_k\}$ and $\{x_u\}$ we have:

$$\{e\} = \left(\begin{Bmatrix} \{x_k\} \\ \{x_u\} \end{Bmatrix} - f\left(\begin{Bmatrix} \{x_k\} \\ \{x_u\} \end{Bmatrix}, \{w\} \right) \right)^2 \tag{4.5}$$

The error vector in Eq. 4.5 can be reduced into a scalar by integrating over the size of the input vector and the number of training examples as follows:

$$E = \left\| \left(\begin{Bmatrix} \{x_k\} \\ \{x_u\} \end{Bmatrix} - f\left(\begin{Bmatrix} \{x_k\} \\ \{x_u\} \end{Bmatrix}, \{w\} \right) \right) \right\| \tag{4.6}$$

Equation 4.6 is called the missing data estimation equation. To estimate the missing input values, Eq. 4.6 is minimized. The interrelationships that exist in the data are thus stored in the function f and the network weights $\{w\}$. There are many types of optimization methods that can be applied to accomplish this task and in this chapter genetic algorithm is used. In this chapter we apply the multi-layer perceptron to model the auto-associative memory network. The missing data estimation model explained in this section is illustrated in Fig. 4.2.

Fig. 4.2 Missing data estimation model (Marwala 2009)

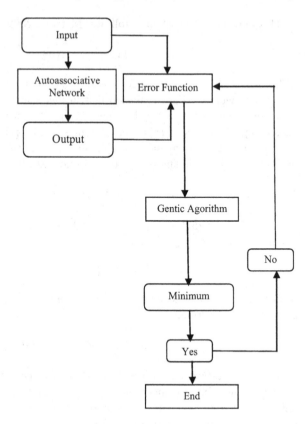

4.3   Multi-layer Perceptron Neural Network

A multi-layer perceptron neural network is a computer-based information processing artificial intelligence machine which is derived from the way biological nervous systems, like the human brain process information (Haykin 1999; Marwala 2007). It is an a powerful artificial machine that has found successful utilization in mechanical engineering (Marwala and Hunt 1999; Marwala 2001, 2010; Mohamed et al. 2006), civil engineering (Marwala 2000), aerospace engineering (Marwala 2003, 2012; Mthembu et al. 2011), biomedical engineering (Mohamed et al. 2006; Marwala 2007; Rusell et al. 2012), financial services (Duma et al. 2012) and political science (Marwala and Lagazio 2011). In addition to neural network being viewed with the cognitive sciences lenses, neural network can also be considered to be a generalized regression model that is capable of modelling linear and non-linear data.

Artificial neural network consists of four main parts (Haykin 1999; Marwala 2013, 2014) which are the processing units u_j, where each u_j has a certain activation level $a_j(t)$ at any point in time, weighted interconnections between the various processing units which determine how the activation of one unit leads to input for another unit, an activation rule which acts on the set of input signals at a unit to produce a new output signal, and a learning rule that specifies how to adjust the weights for a given input/output pair (Haykin 1999; Freeman and Skapura 1991; Marwala 2013).

Because of their capability to make sense of complex data, neural networks can be used to identify patterns and complex trends cannot be easily identified by many mathematical techniques (Hassoun 1995; Yoon and Peterson 1990; Lunga and Marwala 2006).

The architecture of neural processing units and their interconnections in a multi-layer perceptron neural network have significant impact on the processing capabilities of a neural network (Haykin 1999). As a result, there are a number of different artchitectures of how the data flow between the input to the hidden layer and then to the output layer. The multi-layer perceptron (MLP) neural network is one such architecture and is a feed-forward neural network model that approximates a relationship between sets of input data and a set of applicable output. Its basic unit is the standard linear perceptron and it uses three or more layers of neurons (nodes) with non-linear activation functions to effect the algorithmic structure and because of this complexity it is thus more powerful than the ordinary perceptron. The reason for this is because it can classify data that is not linearly separable or separable by a hyperplane. The multi-layered perceptron neural network has been used to model many complex systems in various disciplines (Marwala 2014; Yeh et al. 2014; Xiao et al. 2013; Khong et. al. 2013; Cho 2013).

The multi-layer perceptron neural network consists of multiple layers of computational units which are interconnected in a feed-forward matter and each neuron in one layer is linked to the neurons of the next layer (Haykin 1999; Hassoun 1995). A NETLAB® toolbox that runs in MATLAB® is used to implement the MLP neural network and in this chapter to build a two-layered multi-layer perceptron neural network architecture and this is able to approximate data of arbitrary complexity (Nabney 2001; Haykin 1999). The MLP neural network can be mathematically described as follows (Bishop 1995):

$$y_h = f_{outer} \left(\sum_{j=1}^{M} w_{hj}^{(2)} f_{inner} \left(\sum_{i=1}^{d} w_{ji}^{(1)} x_i + w_{j0}^{(1)} \right) + w_{h0}^{(2)} \right) \tag{4.7}$$

In Eq. 4.7 $w_{ji}^{(1)}$ and $w_{ji}^{(2)}$ indicate weights in the first and second layers, respectively, going from input i to hidden unit j, M is the number of hidden units, d is the number of output units while $w_{j0}^{(1)}$ and $w_{k0}^{(2)}$ are the free parameters that indicate the biases for the hidden unit j and the output unit h. These free parameters are a system that makes the model actually internalize and interpret the data. The parameter $f_{outer}(\bullet)$ is a logistic function while f_{inner} is the hyperbolic tangent function. The choice of these parameters allows the model to be able to model linear and non-linear data of any order. The logistic function is written as follows (Bishop 1995):

$$f_{outer}(v) = \frac{1}{1+e^{-v}} \tag{4.8}$$

The logistic activation function maps the interval $(-\infty, \infty)$ onto a $(0, 1)$ interval and can be approximated by a linear function provided the magnitude of v is small. The hyperbolic tangent function is:

$$f_{inner}(v) = \tanh(v) \tag{4.9}$$

4.3.1 Training the Multi-Layered Perceptron

The network weights in Eq. 4.7 are identified by training the neural network using the given data. This training procedure results in a model that manifests the rules and inter-relationships that govern the data. An objective function that represents some distance between the model prediction and the observed target data with the free parameters as unknown variables must be selected for optimization in order to train the neural network. This is achieved by minimizing this objective function thereby identifying the free parameters known as weights in Eq. 4.7 given the training data. An objective function is a mathematical representation of the principal objective of the problem. In this chapter, the main objective, and this is used to formulate a cost function, is to identify a set of neural network weights that map the input variables onto themselves and this is called the autoassociative network (Marwala 2009). If the training set $D = \{x_k, y_k\}_{k=1}^{N}$ is used and given that the targets y are sampled independently given the inputs x_k and the weight parameters, w_{kj}, the objective function, E, may be written using the sum-of-squares of errors objective function as follows (Bishop 1995):

$$\begin{aligned} E &= \sum_{n=1}^{N}\sum_{k=1}^{K}\{t_{nk} - y_{nk}\}^2 \\ &= \sum_{n=1}^{N}\sum_{k=1}^{K}\{t_{nk} - y_{nk}(\{x\},\{w\})\}^2 \end{aligned} \tag{4.10}$$

In Eq. 4.10, n is the index for the training example, k is the index for the output units, $\{x\}$ is the input vector and $\{w\}$ is the weight vector. Prior to neural network training, the network architecture is created by selecting the number of hidden units, M. If M is too small, the neural network is not adequately flexible and generalizes poorly the data due to high bias. If M is too large, the neural network is too complex and generalizes poorly because of over-fitting due to high variance. The MLP neural network is trained using back-propagation which is a method for finding the derivatives of the error in Eq. 4.10 with respect to the network weights.

4.3.2 Back-Propagation Method

To identify the neural network weights given the training data, an optimization method is implemented. Using a method which is based on the Newton-Raphson method, the neural network weights are identified using the following iterative method (Werbos 1974):

$$\{w\}_{i+1} = \{w\}_i - \eta \frac{\partial E}{\partial \{w\}}(\{w\}_i) \tag{4.11}$$

In Eq. 4.11 the parameter η is the learning rate while $\{\}$ represents a vector. The minimization of the objective function E is achieved by calculating the derivative of the errors in Eq 4.11 with respect to the network weight. The derivative of the error is calculated with respect to the weight which connects the hidden layer to the output layer and may be written using the chain rule as follows (Bishop 1995):

$$
\begin{aligned}
\frac{\partial E}{\partial w_{kj}} &= \frac{\partial E}{\partial a_k} \frac{\partial a_k}{\partial w_{kj}} \\
&= \frac{\partial E}{\partial y_k} \frac{\partial y_k}{\partial a_k} \frac{\partial a_k}{\partial w_{kj}} \\
&= \sum_n f'_{outer}(a_k) \frac{\partial E}{\partial y_{nk}} z_j
\end{aligned}
\tag{4.12}
$$

In Eq. 4.12 $z_j = f_{inner}(a_j)$ and $a_k = \sum_{j=0}^{M} w_{kj}^{(2)} y_j$. The derivative of the error with respect to weight which connects the hidden layer to the output layer may be written using the chain rule as follows (Bishop 1995):

$$\frac{\partial E}{\partial w_{kj}} = \frac{\partial E}{\partial a_k} \frac{\partial a_k}{\partial w_{kj}} = \sum_n f'_{inner}(a_j) \sum_k w_{kj} f'_{outer}(a_k) \frac{\partial E}{\partial y_{nk}} \tag{4.13}$$

In Eq. 4.13, $a_j = \sum_{i=1}^{d} w_{ji}^{(1)} x_i$. The derivative of the sum of square cost function in Eq. 4.13 may be written as follows:

$$\frac{\partial E}{\partial y_{nk}} = t_{nk} - y_{nk} \tag{4.14}$$

while that of the hyperbolic tangent function is:

$$f'_{inner}(a_j) = \text{sech}^2(a_j) \tag{4.15}$$

Now that it has been determined how to calculate the gradient of the error with respect to the network weights using back-propagation algorithms, then Eq. 4.11 is used to update the network weights using an optimization process until some pre-defined stopping condition is achieved. If the learning rate in Eq. 4.11 is fixed then this is known as the steepest descent optimization method (Robbins and Monro 1951). The steepest descent method is not computationally efficient and, therefore, an improved method needs to be found and in this chapter the scaled conjugate method is implemented (Møller 1993).

4.3.3   Scaled Conjugate Method

The method for identifying free parameters (network weights) from the data is by the use of some non-linear optimization technique (Mordecai 2003) and in this chapter the scaled conjugate method. Before scaled conjugate method is described, it is important to comprehend how conjugate gradient method works. In Eq. 4.11, the weight vector which gives the minimum error is obtained from taking successive steps in the neural network weight space until a prescribed stopping criterion is obtained. In order to discuss the scaled conjugate gradient method, we first discuss the gradient descent method (Hestenes and Stiefel 1952). In the gradient descent technique, the step size is defined as $-\eta \partial E / \partial w$, where the parameter η is the step size which is often called the learning rate and the gradient of the error are calculated using the back-propagation method described in the previous section.

If the step size is small enough, the value of error decreases at each successive step until a minimum value for the error between the model prediction and training target data is realized. This approach is, however, computationally expensive when compared to other methods. For the conjugate gradient method, the quadratic function of error is minimized, at each iteration, over a progressively expanding linear vector space that includes the global minimum of the error (Luenberger 1984; Fletcher 1987; Bertsekas 1995). The conjugate gradient method is applied by following these steps as was explained by Haykin (1999) and Marwala (2009):

1. Select the initial weight vector $\{w\}_0$.
2. Calculate the gradient vector $\dfrac{\partial E}{\partial \{w\}}(\{w\}_0)$.
3. At each step n use the line search to find $\eta(n)$ that minimizes $E(\eta)$ representing the cost function expressed in terms of η for fixed values of w and $-\dfrac{\partial E}{\partial \{w\}}(\{w\}_n)$.
4. Check that the Euclidean norm of the vector $-\dfrac{\partial E}{\partial w}(\{w\}_n)$ is sufficiently less than that of $-\dfrac{\partial E}{\partial w}(\{w\}_0)$.
5. Update the weight vector using Eq. 4.11.
6. For w_{n+1} compute the updated gradient $\dfrac{\partial E}{\partial \{w\}}(\{w\}_{n+1})$.
7. Use Polak-Ribiére method to calculate:

$$\beta(n+1) = \frac{\nabla E(\{w\}_{n+1})^T (\nabla E(\{w\}_{n+1}) - \nabla E(\{w\}_n)))}{\nabla E(\{w\}_n)^T \nabla E(\{w\}_n)}$$

8. Update the direction vector

$$\frac{\partial E}{\partial \{w\}}(\{w\}_{n+2}) = \frac{\partial E}{\partial \{w\}}(\{w\}_{n+1}) - \beta(n+1)\frac{\partial E}{\partial \{w\}}(\{w\}_n).$$

9. Set $n=n+1$ and go back to Step 3.
10. Stop when the following condition is satisfied:

$$\frac{\partial E}{\partial \{w\}}(\{w\}_{n+2}) = \varepsilon \frac{\partial E}{\partial \{w\}}(\{w\}_{n+1}) \text{ where } \varepsilon \text{ isasmallnumber.}$$

The scaled conjugate gradient method differs from conjugate gradient method in that it does not involve the line search described in Step 3 in the previous section. The step-size (see Step 3) is calculated directly by using the following formula (Møller 1993):

$$\eta(n) = 2\left(\frac{\eta(n) - \left(\frac{\partial E(n)}{\partial \{w\}}(n)\right)^T H(n)\left(\frac{\partial E(n)}{\partial \{w\}}(n)\right) + \eta(n)\left\|\left(\frac{\partial E(n)}{\partial \{w\}}(n)\right)\right\|^2}{\left\|\left(\frac{\partial E(n)}{\partial \{w\}}(n)\right)\right\|}\right)^2 \quad (4.16)$$

where H is the Hessian of the gradient. The scaled conjugate gradient method is used because it has been found to solve the optimization problem encountered when training an MLP network more computationally efficient than the gradient descent and conjugate gradient methods (Bishop 1995).

The missing data can be estimated using Eq. 4.6 which is thus adapted to include the multi-layer perceptron in Eq. 4.17.

$$E = \left\| \left(\begin{Bmatrix} \{x_k\} \\ \{x_u\} \end{Bmatrix} - \left\{ f\left(\sum_{j=1}^{M} w_{hj}^{(2)} f_{inner}\left(\sum_{i=1}^{d_1} w_{ji}^{(1)} x_{k_i} + \sum_{i=1}^{d_2} w_{ji}^{(1)} x_{u_i} + w_{j0}^{(1)} + w_{j0}^{(1)} \right) + w_{h0}^{(2)} \right) \right\} \right) \right\|$$

$$(4.17)$$

Here h is the index of the vector of output, u is for unknown and k is known, d_1 is the number of known data while d_2 is the number of known data. In Eq. 4.17, the neural network weights are known and some component of x are known (x_{k_i}) while others are unknown (x_{u_i}) so the missing data estimation problem is to identify x_{u_i} by minimizing Eq. 4.17 which is achieved by using genetic algorithm which is the subject of the next section.

4.4 Genetic Algorithm

Genetic algorithm (GA) which is used to identify missing data by minimizing the error eq. 4.17 is an evolutionary optimization algorithm which is derived from the principles of evolutionary biology. GA is particularly useful on identifying approximate solutions to difficult optimization problems such as the travelling salesman problem (Goldberg 1989; Mitchell 1996; Forrest 1996; Vose 1999; Marwala 2010; Xing and Gao 2013). It uses biologically derived techniques such as inheritance, mutation, natural selection and recombination to approximate an optimal solution to difficult problems (Banzhaf et al. 1998; Marwala 2009). Genetic algorithms view learning as a competition among a population of evolving candidate problem solutions. A fitness function (objective function) evaluates each solution to decide whether it will contribute to the next generation of solutions. Through operations analogous to gene transfer in sexual reproduction, the algorithm creates a new population of candidate solutions (Banzhaf et al. 1998; Goldberg 1989).

The three most important aspects of using genetic algorithms are (Michalewicz 1996; Houck et al. 1995) definition of the objective function, definition and implementation of the genetic representation, and definition as well as implementation of the genetic operators. GAs have been proven to be successful in optimization problems such as wire routing, scheduling, adaptive control, game playing, cognitive modeling, transportation problems, traveling salesman problems, optimal control problems and database query optimization (Michalewicz 1996; Falkenauer 1997; Pendharkar and Rodger 1999; Marwala 2001, 2002, 2007; Marwala and Chakraverty 2006; Crossingham and Marwala 2007; Hulley and Marwala 2007).

The MATLAB® implementation of genetic algorithm described in Houck et al. (1995) is used to implement genetic algorithm in this chapter. After executing the program with different genetic operators, optimal operators that give the best results are selected to be used in conducting the experiment. To implement genetic algorithms the following steps are followed: initialization, selection, reproduction and termination.

4.4.1 Initialization

Initially a number of possible individual solutions are randomly generated to form an initial population. This initial population should be sampled randomly over a large representation of the solution space. The size of the population should depend on the nature of the problem which is determined by the number of variables.

4.4.2 Selection

For every generation, a choice of the proportion of the existing population is chosen to breed a new population (Koza 1992). This selection is conducted using the fitness-based process, where solutions that are fitter, as measured by the error function given in Eq. 4.17, are given a higher probability of being chosen. Some selection techniques rank the fitness of each solution and choose the best solutions while other methods rank a randomly chosen sample of the population for computational efficiency.

The majority of the selection functions tend to be stochastic in nature and thus are designed in such a way that a selection process is conducted on a small proportion of less fit solutions. This ensures that diversity of the population of possible solutions is maintained at high level and, therefore, avoids convergence on poor and incorrect solutions. There are many selection methods and these include roulette wheel selection, which is used in this chapter.

Roulette-wheel selection is a genetic operator utilized for selecting potentially useful solutions in genetic algorithm optimization process. In this method, each possible procedure is assigned the fitness function which is utilized to map the probability of selection with each individual solution. Suppose the fitness f_i is of individual i in the population, the probability that this individual is selected is:

$$p_i = f_i \bigg/ \sum_{j=1}^{N} f_j \tag{4.18}$$

In Eq. 4.18 N, is the total population size. This process ensures that candidate solutions with a higher fitness have a lower probability that they may be eliminated than those with lower fitness. By the same token, solutions with low fitness have a low probability of surviving the selection process. The advantage of this is that even though a solution may have low fitness, it may still contain some components which may be useful in the future.

4.4.3 Reproduction, Crossover and Mutation

Reproduction generates subsequent population of solutions from those selected through genetic operators which are crossover and mutation. The crossover operator mixes genetic information in the population by cutting pairs of chromosomes at random points along their length and exchanging over the cut sections. This has a potential of joining successful operators together. Crossover occurs with a certain probability (Kaya 2011). In many natural systems, the probability of crossover occurring is higher than the probability of mutation occurring. Simple crossover technique is used in this chapter (Goldberg 1989). For simple crossover, one crossover point is selected, binary string from beginning of chromosome to the crossover point is copied from one parent, and the rest is copied from the second parent. For example as described by Marwala (2009), when **11001**011 undergoes simple crossover with 11011**111** it becomes **11001111.**

The mutation operator picks a binary digit of the chromosomes at random and inverts it (Holland 1975). This has a potential of introducing to the population new information. Mutation occurs with a certain probability. In many natural systems, the probability of mutation is low (i.e. less than 1 %). In this chapter, binary mutation is used (Goldberg 1989). When binary mutation is used a number written in binary form is chosen and its value is inverted. For an example: 11001011 may become 11000011.

The processes described result in the subsequent generation of population of solutions that is different from the previous generation and that has average fitness that is higher than the precious generation.

4.4.4 Termination

The process described is repeated until a termination condition has been achieved because either a desired solution is found that satisfies the objective function or a specified number of generations has been reached or the solution's fitness converged or any combinations these. The process described above can be written in pseudo-code algorithmic form as explained by Goldberg (1989) and much later by Marwala (2009):

1. Select the initial population
2. Calculate the fitness of each individual in the population using the error function:
3. Repeat

    a. Choose individuals with higher fitness to reproduce
    b. Generate new population using crossover and mutation to produce offspring
    c. Calculate the fitness of each offspring
    d. Replace low fitness section of population with offspring

4. Repeat until termination

4.5 Application to Antenatal Data

This study is based on the antenatal data which was collected in South African Public hospitals and clinics by the South African government and the details of which were described by Leke et al. (2006), Tim and Marwala (2006), Tim (2007) and Leke (2008). Epidemiology studies the function of host, agent and environment to describe the occurrence and spread of disease. Risk factor epidemiology studies the demographic and social features of individuals and endeavors to identify the factors that puts an individual at risk of getting a disease (Poundstone et al. 2004). In this chapter as was done by Tim (2007) and Leke (2008), the demographic and social features of individuals and their conduct are used to identify the risk of HIV infection (Fee and Krieger 1993). The primary driver of HIV is sexual contact, however, social factors also contribute to the risk of exposure and the probability of transmission. Determining the individual risk factors that lead to HIV allows governments to adjust social conditions minimize HIV infection (Poundstone et al. 2004).

The variables that are gathered in this data are in Table 4.1. These variables are: race, region, age of the mother and the father, education level of the mother, gravidity, parity as well as HIV status (Leke et al. 2006; Tim and Marwala 2006; Tim 2007; Leke 2008). The qualitative variables such as race and region are converted to

Table 4.1 Data from the antenatal dataset

Variable	Type	Range
Region A	Binary	0–1
Region B	Binary	0–1
Region C	Binary	0–1
Age	Integer	15–49
Race: african	Binary	[00]
Race: colored	Binary	[10]
Race: white	Binary	[01]
Race: asian	Binary	[11]
Education	Integer	1–13
Gravidity	Integer	1–12
Parity	Integer	1–12
Age of the father	Integer	15–54
HIV status	Binary	0–1

binary values in order to prevent placing an incorrect importance on these variables had they been coded numerically. The age of mother and father are represented in years. The integer value representing education level represents the highest grade successfully completed, with 13 representing tertiary education. Gravidity is the number of pregnancies, complete or incomplete, experienced by a female, and this variable is represented by an integer between 0 and 12. Parity is the number of times the individual has given birth, (for example, multiple births are counted as one) and this is not the same as gravidity. Both these quantities are important, as they show the reproductive activity as well as the reproductive health state of the women. The HIV status is binary coded, a 1 represents positive status, while a 0 represents negative status.

The MLP neural network which was described in earlier sections was used to construct the autoassociative network. This autoassociative network had 13 inputs, 10 hidden units and 13 outputs. The activation function in the second layer was linear while in the hidden layer was a hyperbolic tangent function. The MLP networks were trained using the scaled conjugate gradient method. 1000 data points, 500 HIV negative and 500 HIV positive were used to train the networks. After the autoassociative neural network model was trained then 500 data points were used to estimate the missing HIV data. This was done using Eq. 4.17 and solving this equation using genetic algorithm. The genetic algorithm had a population of 20 and was run for 40 generations. The results obtained are shown in Table 4.2. In this figure the specificity which is a measure of HIV negative cases that were classified correctly was 62%. The sensitivity which is the proportion of HIV positive cases which were classified correctly was 62% and the overall classification accuracy was 62.5%. Even though these results are not very high they clearly indicate that the missing data estimation technique performs better than random guess (50%) and performs marginally better using a multi-layer perceptron network that was conducted by Tim (2007) who obtained 62%.

Table 4.2 Results obtained
from the missing data
approach based on the multi-
layer perceptron and genetic
algorithm

Measure	Results (%)
Sensitivity (%)	63
Specificity (%)	62
Accuracy (%)	62.5

Conclusions

This chapter describes a missing data estimation method which is based on the multi-layer perceptron autoassociative neural network and genetic algorithm. This technique is used to predict HIV status of a subject given the demographic characteristics. The results indicate that missing data approaches are able to correctly predict the HIV status given the demographic characteristics.

References

Abdella M (2005) The use of genetic algorithms and neural networks to approximate missing data in Database Master Thesis, University of the Witwatersrand

Abdella M, Marwala T (2005) Treatment of missing data using neural networks. Proceedings of the IEEE International Joint Conference on Neural Networks, Montreal, Canada, pp 598–603

Abdella M, Marwala T (2006) The use of genetic algorithms and neural networks to approximate missing data in database. Comput Inf 24:1001–1013

Banzhaf W, Nordin P, Keller R, Francone F (1998) Genetic programming-an introduction: on the automatic evolution of computer programs and its applications, 5th edn. Morgan Kaufmann, California

Bertsekas DP (1995) Nonlinear programming, Athenas Scientific, Belmont. (Marwala T (2014) Causality, correlation and artificial intelligence for rational decision making. World Scientific Publications, Singapore (in press))

Beunckens C, Sotto C, Molenberghs G (2008) A simulation study comparing weighted estimating equations with multiple imputation based estimating equations for longitudinal binary data. Comput Stat Data Anal 52(3):1533–1548

Bishop CM (1995) Neural networks for pattern recognition. Oxford, London

Cho K (2013) Understanding dropout: training multi-layer perceptrons with auxiliary independent stochastic neurons. Lecture notes in Computer Science (including subseries Lecture Notes in Artificial Intelligence and Lecture Notes in Bioinformatics), 8226 LNCS (Part 1), pp 474–481

Crossingham B, Marwala T (2007) Using genetic algorithms to optimise rough set partition sizes for HIV data analysis. Comput Intell 78:245–250

Dempster AP, Laird NM, Rubin DB (1977) Maximum likelihood for incomplete data via the EM algorithm. J R Stat Soc B39:1–38

Dhlamini SM, Nelwamondo FV, Marwala T (2006) Condition monitoring of HV bushings in the presence of missing data using evolutionary computing. Trans Power Syst 1(2):280–287

Donders RT, van der Heijden GJMG, Stijnen T, Moons KGM (2006) Review: a gentle introduction to imputation of missing values. J Clin Epidemiol 59(10):1087–1091

Duma M, Twala B, Nelwamondo F, Marwala T (2012) Predictive modeling with missing data using an automatic relevance determination ensemble: a comparative study. Appl Artif Intell 26:967–984

Falkenauer E (1997) Genetic algorithms and grouping problems. Wiley, Chichester

Faris PD, Ghali WA, Brant R, Norris CM, Galbraith PD, Knudtson ML (2002) Multiple imputation versus data enhancement for dealing with missing data in observational health care outcome analyses. J Clin Epidemiol 55(2):184–191

Fee E, Krieger N (1993) Understanding AIDS: historical interpretations and limits of biomedical individualism. Am J Public Health 83:1477–1488

Fletcher R (1987) Practical methods of optimization, 2nd edn. Wiley, New York

Forrest S (1996) Genetic algorithms. ACM Comput Surv 28:77–80

Freeman J, Skapura D (1991) Neural networks: algorithms, applications and programming techniques. Addison-Wesley, Boston

Gabrys B (2002) Neuro-fuzzy approach to processing inputs with missing values in pattern recognition problems. Int J Approx Reason 30:149–179

Gad AM, Ahmed AS (2006) Analysis of longitudinal data with intermittent missing values using the stochastic EM algorithm. Comput Stat Data Anal 50(10):2702–2714

Goldberg DE (1989) Genetic algorithms in search, optimization, and machine learning. Addison-Wesley, Reading

Harel O (2007) Inferences on missing information under multiple imputation and two-stage multiple imputation. Stat Method 4(1):75–89

Hassoun MH (1995) Fundamentals of artificial neural networks. MIT, Cambridge

Haykin S (1999) Neural networks, 2nd edn. Prentice-Hall, New Jersey

Hestenes MR, Stiefel E (1952) Methods of conjugate gradients for solving linear systems. J Res Nat Bur Stand 6:409–436

Ho P, Silva MCM, Hogg TA (2001) Multiple imputation and maximum likelihood principal component analysis of incomplete multivariate data from a study of the ageing of port. Chemometr Intell Lab 55(1–2):1–11

Holland J (1975) Adaptation in natural and artificial systems. University of Michigan, Ann Arbor

Houck CR, Joines JA, Kay MG (1995) A genetic algorithm for function optimisation: a MATLAB implementation, Tech. Rep. NCSU-IE TR 95-09, North Carolina State University

Huang X, Zhu Q (2002) A pseudo-nearest-neighbour approach for missing data recovery on Gaussian random data sets. Pattern Recognit Lett 23:1613–1622

Hulley G, Marwala T (2007) Genetic algorithm based incremental learning for optimal weight and classifier selection. Am I Phys Ser 952:258–268

Junninen H, Niska H, Tuppurainen K, Ruuskanen J, Kolehmainen M (2007) Methods for imputation of missing values in air quality data sets. Atmos Environ 38(18):2895–2907

Kaya M (2011) The effects of two new crossover operators on genetic algorithm performance. Appl Soft Comput J 11:881–890

Khong LMD, Gale TJ, Jiang D, Olivier JC, Ortiz-Catalan M (2013) Multi-layer perceptron training algorithms for pattern recognition of myoelectric signals. BMEiCON 2013-6th Biomedical Engineering International Conference, art. no. 6687665

Koza J (1992) Genetic programming: on the programming of computers by means of natural selection. MIT, Cambridge

Leke BB (2008) Computational intelligence for modelling HIV. University of the Witwatersrand Ph. D. Thesis

Leke BB, Marwala T, Tim T, Lagazio M (2006) Prediction of HIV Status from Demographic Data Using Neural Networks. Proceedings of the IEEE International Conference on Systems, Man and Cybernetics, Taiwan, pp 2339–2344

Luenberger DG (1984) Linear and nonlinear programming, 2nd edn. Addison-Wesley, Reading

Lunga D, Marwala T (2006) Online forecasting of stock market movement direction using the improved incremental algorithm. Lect Notes Comput Sci 4234:440–449

Marwala T (2000) On damage identification using a committee of neural networks. American society of civil egineers. J Eng Mech 126:43–50

Marwala T (2001) Fault identification using neural networks and vibration data. PhD Thesis, University of Cambridge

Marwala T (2002) Finite element updating using wavelet data and genetic algorithm. American institute of aeronaut and astronaut. J Aircraft 39:709–711

Marwala T (2003) Fault classification using pseudo modal energies and neural networks. Am Inst Aeronaut Astronaut J 41:82–89

Marwala T (2007) Bayesian training of neural network using genetic programming. Pattern Recogn Lett. doi:org/10.1016/j.patrec.2007.034

Marwala T (2009) Computational intelligence for missing data imputation, estimation and management: knowledge optimization techniques. IGI Global, New York

Marwala T (2010) Finite element model updating using computational intelligence techniques. Springer, Heidelberg

Marwala T (2012) Condition monitoring using computational intelligence methods. Springer, Heidelberg

Marwala T (2013) Economic modelling using artificial intelligence methods. Springer, Heidelberg

Marwala T (2014) Causality, correlation and artificial intelligence for rational decision making. World Scientific, Singapore

Marwala T, Chakraverty S (2006) Fault classification in structures with incomplete measured data using autoassociative neural networks and genetic algorithm. Curr Sci 90:542–548

Marwala T, Hunt HEM (1999) Fault identification using finite element models and neural networks. Mech Syst Signal Process 13:475–490

Marwala T, Lagazio M (2011) Militarized conflict modeling using computational intelligence techniques. Springer, Heidelberg

Michalewicz Z (1996) Genetic algorithms + data structures = evolution programs. Springer, New York

Mitchell M (1996) An introduction to genetic algorithms. MIT, Cambridge

Mohamed N, Rubin D, Marwala T (2006) Detection of epileptiform activity in human EEG signals using Bayesian neural networks. Neural Inf Process—Lett Rev 10:1–10

Mohamed S, Tettey T, Marwala T (2006) An extension neural network and genetic algorithm for bearing fault classification. In the Proceedings of the IEEE International Joint Conference on Neural Networks, BC, Canada, pp 7673–7679

Møller AF (1993) A scaled conjugate gradient algorithm for fast supervised learning. Neural Netw 6:525–533

Mordecai A (2003) Nonlinear programming: analysis and methods. Dover New York

Mthembu L, Marwala T, Friswell MI, Adhikari S (2011) Model selection in finite element model updating using the Bayesian evidence statistic. Mechanical Systems and Signal Processing. doi:10.1016/j.ymssp.2011.04.001

Nabney IT (2001) Netlab: algorithms for pattern recognition. Springer, United Kingdom

Nelwamondo FV (2008) Computational Intelligence Techniques for Missing Data Imputation. PhD Thesis, University of the Witwatersrand

Pendharkar PC, Rodger JA (1999) An empirical study of nonbinary genetic algorithm-based neural approaches for classification. Proceedings of the 20th international conference on Information Systems, pp 155–165

Poundstone K, Strathdee S, Celectano D (2004) The social epidemiology of human immunodeficiency virus/acquired Immunodeficiency syndrome. Epidemiol Rev 26:22–35

Reiter JP (2008) Selecting the number of imputed datasets when using multiple imputation for missing data and disclosure limitation. Stat Probabil Lett 78(1):15–20

Robbins H, Monro S (1951) A stochastic approximation method. Ann Math Stat 22:400–407

Rubin DB (1987) Multiple imputation for nonresponse in surveys. Wiley, New York

Russell MJ, Nel A, Marwala T (2012) ARMA analysis of chest X-rays for computer assisted detection of tuberculosis, World Congress on Medical Physics and Biomedical Engineering May 26–31, 2012, Beijing, China. IFMBE Proceedings vol 39, pp 896–899

Sartori N, Salvan A, Thomaseth K (2005) Multiple imputation of missing values in a cancer mortality analysis with estimated exposure dose. Comput Stat Data An 49(3):937–953

Tim T (2007) Predicting HIV status using neural networks and demographic factors. University of the Witwatersrand M.Sc. Master Thesis

Tim TN, Marwala T (2006) Computational Intelligence Methods for Risk Assessment of HIV. In Imaging the Future Medicine, Proceedings of the IFMBE, vol 14, pp 3581–3585, Springer-Verlag, Berlin Heidelberg. Sun I. Kim and Tae Suk Sah (eds), ISBN: 978-3-540-36839-7

Vose MD (1999) The simple genetic algorithm: foundations and theory. MIT, Cambridge

Werbos PJ (1974) Beyond Regression: New Tool for Prediction and Analysis in the Behavioral Sciences. Ph.D. thesis, Harvard University

Xiao W, Pu D, Dong Z, Liu C (2013) The application of optimal weights initialization algorithm based on K-L transfrom in multi-layer perceptron networks. Proceedings of SPIE—The International Society for Optical Engineering, 8878, art. no. 88784O

Xing B, Gao W-J (2013) Innovative computational intelligence: a rough guide to 134 clever algorithms. Springer, Heidelberg

Yeh W-C, Yeh Y-M, Chang P-C, Ke Y-C, Chung V (2014) Forecasting wind power in the Mai Liao Wind Farm based on the multi-layer perceptron artificial neural network model with improved simplified swarm optimization. Int J Electric Power Energy Syst 55:741–748

Yoon Y, Peterson LL (1990) Artificial neural networks: an emerging new technique. Proceedings of the 1990 ACM SIGBDP Conference on Trends and Directions in Expert Systems, pp 417–422

Chapter 5
Rational Counterfactuals and Decision Making: Application to Interstate Conflict

5.1 Introduction

Rational decision making is important for many areas including economics, political science and engineering. Rational decision making involves choosing a course of action which maximizes the net utility. This chapter explores counterfactual thinking in particular and introduces the theory of rational counterfactuals for rational decision making. The idea of using counterfactual thinking for decision making is an old concept that has been explored extensively by many researchers before (Lewis 1973, 1979).

In counterfactual thinking factual statements like: 'Saddam Hussein invaded Kuwait consequently George Bush declared war on Iraq', has a counterfactual: 'Saddam Hussein invaded Kuwait and consequently George Bush declared war on Iraq'. If Saddam Hussein did invade Kuwait then George Bush would not have declared war on Iraq, then there would not have been a causal relation between Saddam invading Kuwait and Bush declaring war on Iraq. Counterfactual thinking has been applied for decision making and is essentially a process of comparing a real and hypothetical thinking and using the difference between these to make decisions.

Howlett and Paulus (2013) applied counterfactual thinking successfully in the problem of patient depression whereas Leach and Patall (2013) applied counterfactual reasoning for decision making on academic major. Celuch and Saxby (2013) applied successfully counterfactual reasoning in ethical decision making in marketing education whereas Simioni et al. (2012) observed that multiple sclerosis decreases explicit counterfactual processing and risk taking in decision making. Fogel and Berry (2010) studied the usefulness of regret and counterfactuals on the relationship between disposition and individual investor decisions.

Johansson and Broström (2011) successfully applied counterfactual thinking in surrogate decision making. In this context incompetent patients have someone make decisions on their behalf by having a surrogate decision maker make the decision that the patient would have made. Daftary-Kapur and Berry (2010) applied counterfactual reasoning on juror punitive damage award decision making while Shaffer (2009) studied decision theory, intelligent planning and counterfactuals to

© Springer International Publishing Switzerland 2014   73
T. Marwala, *Artificial Intelligence Techniques for Rational Decision Making,*
Advanced Information and Knowledge Processing,
DOI 10.1007/978-3-319-11424-8_5

understand the limitation of the fact that Bayesian decision-theoretic framework does not sufficiently explain the causal links between acts, states and outcomes in decision making.

In this chapter, we develop a framework called a rational counterfactual machine which is a computational tool which takes in a factual and gives a counterfactual that is based on optimizing for the desired consequent by identifying an appropriate antecedent. This counterfactual is based on the learning machine and in this chapter we choose the neuro-fuzzy network (Montazer et al. 2010; Talei et al. 2010) and simulated annealing optimization (De Vicente et al. 2003; Dafflon et al. 2009). The rational counterfactual machine is applied to identify the antecedent that will give the consequent which is different from the consequent of the factual, and the example that is used in this chapter is a problem of interstate conflict. This is done in a similar manner as it was done by Marwala and Lagazio (2004, 2011) as well as by Tettey and Marwala (2004). The rational counterfactual machine is applied here to identify the values of antecedent variables *Allies*, *Contingency*, *Distance*, *Major Power*, *Capability*, *Democracy*, as well as *Economic Interdependency* that will give the consequent *Peace* given the factual statement.

This chapter is organized as follows: the next section describes the notion of counterfactuals and then describes the rational counterfactual machine. Then a learning method and in this chapter a neuro-fuzzy method is described. Thereafter, simulated annealing is described followed by the application of the rational counterfactual machine based on neuro-fuzzy model and simulated annealing in the problem of interstate conflict.

5.2   Counterfactuals

Counterfactual thinking has been around for a very long time. Some of the thinkers who have dealt with the concept of counterfactuals include Hume (1748), Mill (1843), Hegel's dialectic concept of thesis (i.e. factual), antithesis (i.e. counterfactual) and synthesis (Hegel 1874) and Marx (1873). Counterfactual can be understood by breaking this word into two parts *counter* and *factual*. Factual is an event that has happened for example: Saddam Hussein invaded Kuwait and consequently George Bush declared war on Iraq. *Counter* means the opposite of and in the light of the factual above: If Saddam Hussein did not invade Kuwait and consequently George Bush would not have declared war on Iraq. Of course counterfactual can be an imaginary concept and, therefore, the fundamental question that needs to be asked is: How do we know what would have happened if something did not happen? This book addresses classes of problems where it is possible to estimate what might have happened and this procedure which we call a rational counterfactual machine is implemented via artificial intelligence techniques.

There are different types of counterfactuals and these include self/other as well as well as additive/subtractive. Additive and subtractive counterfactual is the case were the antecedent is either increased or decreased. One example of such will

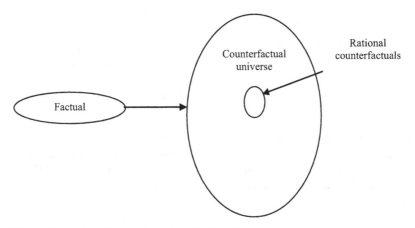

Fig 5.1. An illustration of a transformation of a factual into a counterfactual

include: He drank alcohol moderately and consequently he did not fall sick. The counterfactual of this statement might be: He drank a lot alcohol and consequently he fell sick. The 'a lot' adds to the antecedent in the counterfactual.

There are a number of theories that have been proposed to understand counterfactuals and these include norm and functional theories (Birke et al. 2011; Roese 1997). As described by Kahneman and Miller (1986) norm theory comprises a pairwise evaluation between a cognitive standard and an experiential outcome. Functional theory entails looking at how a counterfactual theory and its processes benefit people. Rational counterfactuals can be viewed as an example of the functional theory of counterfactuals.

Figure 5.1 indicates a factual and its transformation into a counterfactual. It indicates that in the universe of counterfactuals that correspond to the factual there are many if not infinite number of counterfactuals. Suppose we have a factual: Mandela opposed apartheid and consequently he went to jail for 27 years. Its counterfactual can be: If Mandela did not oppose apartheid then he would not have gone to jail or If Mandela opposed apartheid gently he would not have gone to jail or If Mandela opposed apartheid peacefully he would not have gone to jail. It is clear that there are multiple ways in which one can formulate counterfactuals for a given factual.

There a number of ways in which counterfactuals can be stated and this involves structural equations (Woodward 2003; Woodward and Hitchcock 2003). In the structural equation approach an expression can be stated as a counterfactual as follows:

$$y = f(x_1, x_2, \ldots, x_n) \tag{5.1}$$

This expression can be read as: if it is the case that $x_1 = X_1, x_2 = X_2, \ldots, x_n = X_n$ then it will be the case that $y = f(X_1, X_2, \ldots, X_n)$. The usefulness of this approach will be apparent later in the chapter when it is applied to modelling interstate conflict.

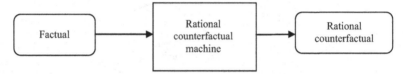

Fig. 5.2 An illustration of a rational counterfactual machine

Figure 5.1 states that within the counterfactual universe there are group of counterfactuals that are called rational counterfactuals which are counterfactuals that are designed to maximize the attainment of particular consequences and these are called rational counterfactuals and are the subject of the next section.

5.3 Rational Counterfactual Machine

Now that we have discussed the concept of counterfactual, this section describes the concept of rational counterfactual and the corresponding machine for creating this concept. As shown in Fig. 5.1, rational counterfactuals are those counterfactuals in the counterfactual universe corresponding to a given factual that maximizes a particular goal. There is a statement attributed to Karl Marx that states: "*The aim of a revolutionary is not merely to understand the world but to actually change it*". In this chapter we, therefore, use counterfactual theory to solve practical problems and this is called the functional theory to counterfactual thinking. In this chapter we also build what is known as a counterfactual machine, which is a computational system which gives a rational counterfactual whenever it is presented with a factual and a given problem domain. An illustration of a rational counterfactual machine is given in Fig. 5.2. This figure shows that there are three objects in a rational counterfactual machine and these are the factual which is antecedent into the rational counterfactual machine to give a rational counterfactual.

The rational counterfactual machine consists of a model that describes the structure and rules that define the problem at hand and a feedback which links the consequent (outcome) of the model and the antecedent. This model is shown in Fig. 5.3.

In this chapter, we apply the problem of interstate conflict to illustrate the concept of rational counterfactuals. In this regard, we use neuro-fuzzy model to construct a factual relationship between the antecedent and the consequent. Then to identify the antecedent given the desired consequent and an optimization method. The objective function of the optimization problem is (Marwala and Lagazio 2011):

$$E = \sum (y - t_d)^2 \tag{5.2}$$

Here, E is the error, y is the neuro-fuzzy consequent and t_d is the desired target consequent. Equation 5.2 is solved using simulated annealing (Marwala and Lagazio

Fig. 5.3 An illustration of
a rational counterfactual
machine

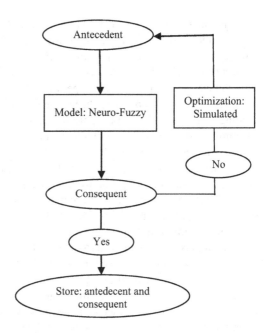

2011). Equation 5.2 allows one to be able to construct a rational counterfactual which relates the identified antecedent (using Eq. 5.2) and the desired consequence. The next section describes the neuro-fuzzy model followed by simulated annealing.

5.4   Neuro-Fuzzy Model

5.4.1   Basic Fuzzy Logic Theory

Fuzzy logic is a procedure for relating an antecedent to a consequent using linguistic rules entailing the if-then statements (Bih 2006; von Altrock 1995; Biacino and Gerla 2002; Cox 1994). It consists of four objects: fuzzy sets, membership functions, fuzzy logic operators and fuzzy rules. In classical set theory a set is either an element of or is not an element of a set (Devlin 1993; Ferreirós 1999; Johnson 1972). Accordingly, it is possible to explain if an object is an element of a set because of clear boundaries. Classically a characteristic function of a set has a value of 1 if the object is an element of that set and a value of zero if the object doesn't belong to the set (Cantor 1874). In this regard there is no middle road in terms of membership or non-membership of a set.

Fuzzy logic offers a flexible representation of sets of objects by introducing a concept of a fuzzy set which does not have as clear cut boundaries as a classical set and the objects have a degree of membership to a particular set (Hájek 1995; Halpern 2003; Wright and Marwala 2006; Hájek 1998).

A membership function defines the degree that an object is a member of a set. The membership function is a curve that maps the antecedent space variable to a number between 0 and 1, representing the degree to which an object is a member of a set (Klir and Folger 1988; Klir et al. 1997; Klir and Yuan 1995). A membership function can be a curve of any shape. For example, if we are studying the problem of height, there would be two subsets one for tall and one for short that overlap. Accordingly, a person can have a partial membership in each of these sets, therefore, determining the degree to which the person is both tall and short.

Logical operators are applied to create new fuzzy sets from the existing fuzzy sets. Classical set theory offers three basic operators for logical expressions to be defined: intersection, union and the complement (Kosko 1993; Kosko and Isaka 1993). These operators are also applied in fuzzy logic and have been adapted to deal with partial memberships. The intersection (AND operator) of two fuzzy sets is given by a minimum operation and the union (OR operator) of two fuzzy sets is given by a maximum operation (Novák 1989, 2005; Novák et al. 1999). These logical operators are applied to determine the final consequent fuzzy set.

Using fuzzy rules conditional statements which are applied to model the antecedent-consequent relationships of the system expressed in natural language are created. These linguistic rules which are expressed in the *if-then statements* use logical operators and membership functions to infer a consequent. A characteristic of fuzzy logic are linguistic variables which use words or sentences as their values instead of numbers (Zadeh 1965; Zemankova-Leech 1983; Zimmermann 2001; Lagazio and Marwala 2011). As described by Marwala and Lagazio (2011), linguistic variable *weight* can be assigned the following term set {*very fat, fat, medium, thin, very thin*} and then a fuzzy rule is of the form:

$$\text{if } x \text{ is } A \text{ then } y \text{ is } B \tag{5.3}$$

where A and B are fuzzy sets defined for the antecedent and consequent space respectively. Both x and y are linguistic variables, while A and B are linguistic values expressed by applying membership functions. Each rule consists of two parts: the antecedent and the consequent. The antecedent is the component of the rule falling between the *if-then* and maps the antecedent x to the fuzzy set A, applying a membership function. The consequent is the component of the rule after the then, and maps the consequent y to a membership function. The antecedent membership values act like weighting factors to determine their influence on the fuzzy consequent sets. A fuzzy system consists of a list of these if-then rules which are evaluated in parallel. The antecedent can have more than one linguistic variable, these antecedents are combined by applying the AND operator.

Each of the rules is evaluated for an antecedent set, and corresponding consequent for the rule obtained. If an antecedent corresponds to two linguistic variable values then the rules associated with both these values will be evaluated. Also, the rest of the rules will be evaluated, however, they will not have an effect on the final result as the linguistic variable will have a value of zero. Therefore, if the antecedent is true to some degree, the consequent will have to be true to some degree (Zadeh

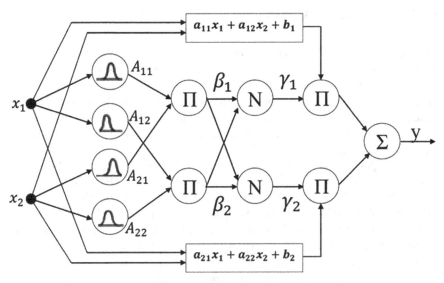

Fig. 5.4 An example of a two-antecedent first order Takagi-Sugeno fuzzy model

1965). The degree of each linguistic consequent value is then computed by performing a combined logical sum for each membership function (Zadeh 1965) after which all the combined sums for a specific linguistic variable can be combined. These last steps involve the use of an inference method which maps the result onto a consequent membership function (Zadeh 1965; Mamdani 1974).

Finally de-fuzzification process is performed where a single numeric consequent is produced. One method of estimating the degree of each linguistic consequent value is to take the maximum of all rules describing this linguistic consequent value, and the consequent is taken as the center of gravity of the area under the effected part of the consequent membership function. There are other inference methods such as averaging and sum of mean of squares. The application of a series of fuzzy rules, and inference methods to produce a de-fuzzified consequent is called a Fuzzy Inference System (FIS).

There are several other types of fuzzy inference systems which vary according to the fuzzy reasoning and the form of the *if-then* statements applied. One of these methods is the Takagi-Sugeno-Kang neuro-fuzzy method (Takagi and Sugeno 1985; Araujo 2008). This technique is similar to the Mamdani approach described above except that the consequent part is of a different form and, as a result, the de-fuzzification procedure is different. The *if-then* statement of a Sugeno fuzzy system expresses the consequent of each rule as a function of the antecedent variables and has the form (Sugeno and Kang 1988; Sugeno 1985):

$$\textit{if } x \textit{ is } A \textbf{ AND } y \textit{ is } B \textbf{ then } z = f(x, y) \tag{5.4}$$

5.4.2   *Neuro-fuzzy Models*

Neuro-fuzzy is a combination of neural networks and fuzzy logic (Jang 1993; Sugeno and Kang 1988; Sugeno 1985; Tettey and Marwala 2007) thereby combining the human-like reasoning style of fuzzy systems with the learning of neural networks and creating a universal approximator with an interpretable *if-then* rules. The advantage of neuro-fuzzy systems is its ease of interpretability as well as inference accuracy. There are different types of neuro-fuzzy models and in this chapter we focus on the Takagi-Sugeno neuro-fuzzy model and the Takagi-Sugeno-Kang neuro-fuzzy model (Takagi and Sugeno 1985; Jang 1997; Araujo 2008). Neuro-fuzzy modeling has been successfully applied in control systems (Hsu and Lin 2009; Iplikci 2010; Cano-Izquierdo et al. 2010), in character recognition (Montazer et al. 2010); to predict solubility of gases in polystyrene (Khajeh and Modarress 2010); to predict tip speed ratio in wind turbines (Ata and Kocyigit 2010) and to predict the moment capacity of Ferro-cement members (Mashrei et al. 2010).

The Takagi-Sugeno neuro-fuzzy model the fuzzy proposition is the antecedent whereas the consequent function is an affine linear function of the antecedent and this can be expressed as follows (Takagi and Sugeno 1985):

$$R_i : \text{If } x \text{ is } A_i \text{ then } y_i = a_i^T x + b_i \tag{5.5}$$

here R_i is the ith fuzzy rule, x is the antecedent vector, A_i is a fuzzy set, a_i is the consequence parameter vector, b_i is a scalar offset and $i = 1, 2, ..., K$. The parameter K is the number of rules in the fuzzy model. Too few rules in the fuzzy model compromises the accuracy of the fuzzy system whereas too many rules gives an unnecessarily complex model and compromises the integrity of the model (Sentes et al. 1998). Optimum number of rules is empirically identified in this chapter using a cross-validation (Bishop 1995). The final antecedent values in the model describe the fuzzy regions in the antecedent space in which the consequent functions are valid. The neuro-fuzzy system applied in this chapter is shown in Fig. 5.4.

The first step in any inference procedure is the partitioning of the antecedent space in order to form the antecedents of the fuzzy rules. The shapes of the membership functions of the antecedents can be chosen to be Gaussian or triangular. The Gaussian membership function of the following form is chosen in this chapter (Zadeh 1965).

$$\mu^i(x) = \prod_{j=1}^{n} e^{-\frac{\left(x_j - c_j^i\right)^2}{\left(b_j^i\right)^2}} \tag{5.6}$$

Here, μ^i is the combined antecedent value for the ith rule, n is the number of antecedents belonging to the ith rule, c is the center of the Gaussian function and b describes the variance of the Gaussian membership function.

The consequent function in the Takagi-Sugeno neuro-fuzzy model can either be constant or linear. In this chapter, a linear consequent function is chosen and is expressed as follows (Babuska 1991; Babuska and Verbruggen 2003):

$$y_i = \sum_{j=1}^{n} p_{ij} x_j + p_{i0} \tag{5.7}$$

where p_{ij} is the jth parameter of the ith fuzzy rule. If a constant is applied as the consequent function, i.e. $y_i = p_i$, the zero-order Takagi-Sugeno neuro-fuzzy model becomes a special case of the Mamdani inference system (Mamdani 1974). The consequent y of the entire inference system is computed by taking a weighted average of the individual rules' contributions as shown in (Babuska and Verbruggen 2003):

$$y = \frac{\sum_{i=1}^{K} \beta_i(x) y_i}{\sum_{i=1}^{K} \beta_i(x)}$$
$$= \frac{\sum_{i=1}^{K} \beta_i(x)(a_i^T x + b_i)}{\sum_{i=1}^{K} \beta_i(x)} \tag{5.8}$$

where $\beta_i(x)$ is the activation of the ith rule. The $\beta_i(x)$ can be a complicated expression but, in our work, it will be equivalent to the degree of fulfilment of the ith rule. The parameters a_i are then approximate models of the non-linear system under consideration (Babuska 1991).

When setting up a fuzzy rule-based system we are required to optimize parameters such as membership functions and consequent parameters. In order to optimize these parameters, the fuzzy system relies on training algorithms inherited from artificial neural networks such as gradient descent-based learning. It is for this reason that they are referred to as neuro-fuzzy models. There are two approaches to training neuro-fuzzy models (Babuska and Verbruggen 2003):

1. Fuzzy rules may be extracted from expert knowledge and applied to create an initial model. The parameters of the model can then be fine-tuned using data collected from the operational system being modelled.
2. The number of rules can be determined from collected numerical data using a model selection technique. The parameters of the model are also optimized using the existing data.

The major motivation for using a Takagi-Sugeno neuro-fuzzy model in this work is that, not only is it suitable for data-driven identification, but it is also considered to be a gray box. Unlike other computational intelligence methods, once optimized, it is possible to extract information which allows one to understand the process being modeled.

5.5 Simulated Annealing (SA)

Simulated Annealing is a Monte Carlo based optimization technique inspired by the annealing process of metals freezing. Monte Carlo method is a computational method that uses repeated random sampling to calculate a result (Mathe and Novak 2007; Akhmatskaya et al. 2009; Ratick and Schwarz 2009; Marwala 2010). Monte Carlo method's reliance on repeated computation and random or pseudo-random numbers necessitates the use of computers. Monte Carlo method is usually applied when it is not possible or not feasible to compute an exact solution using a deterministic algorithm. In annealing process, metal is heated until it is melted and then its temperature is slowly reduced in such a way that the metal, at any given time, is approximately in thermodynamic equilibrium. As the temperature of the object is dropped, the system becomes more ordered and approaches a *frozen* state at $T=0$. If the cooling process is conducted inadequately or the initial temperature of the object is not sufficiently high, the system may become quenched, forming defects or freezing out in meta-stable states. This indicates that the system is trapped in a local minimum energy state.

The process that is followed to simulate the annealing process was first formulated by Metropolis et al. (1953) and it comprises selecting an initial state and temperature, and keeping the temperature constant, perturbing the initial structure and estimating the error of the new state using Eq. 5.2. If the new error is lower than the old error then the new state is accepted, otherwise if the contrary is the case, then this state is accepted with a low probability.

Simulated annealing substitutes a current solution by a random solution with a probability that depends on the difference between the corresponding objective function values and the temperature. As the temperature decreases and approaches zero there is less random changes in the solution. Simulated annealing identifies the global optimum but this can require an infinite amount of time to realize this. The probability of accepting the reversal is given by Boltzmann's equation as follows (Bryan et al. 2006; Marwala 2010):

$$P(\Delta E) = \frac{1}{Z} \exp\left(-\frac{\Delta E}{T}\right) \tag{5.9}$$

Here ΔE is the difference in error between the old and new states. T is the temperature of the system; Z is a normalization factor that ensures that when the probability function is integrated to infinity it becomes 1. The rate at which the temperature decreases depends on the cooling schedule selected. There are many different temperature schedules. SA has been applied extensively to optimally design concrete frames (Paya-Zaforteza et al. 2009), optimal synthesis of distillation (Weizhong et al. 2008), finite element updating (Marwala 2010), missing data estimation (Marwala 2009), damage detection (He and Hwang 2006) as well as vibration control (Moita et al. 2006).

On using simulated annealing several parameters and choices need to be specified. These are (Kirkpatrick et al. 1983; Marwala 2010):

- The state space, which is defined in this chapter as a choice of a set of variables that makes a model give a desired consequent;
- The objective function described in Eq. 5.2.
- The candidate generator mechanism which is a random number generator that ensures that antecedents are chosen and a random walk process is used;
- The acceptance probability function, which is a process through which a set of antecedent gives the desired consequent; and
- The annealing temperature schedule.

The choice of these parameters has influential significances on the effectiveness of simulated annealing identifying an optimal solution. Nevertheless, there is no optimal modus that can be employed to select these parameters and there is no methodical technique for optimally selecting these parameters for a given problem. Accordingly, the choice of these parameters is subjective and a technique of trial and error is generally applied.

When simulated annealing is applied, a random walk process is undertaken for a given temperature and it involves moving from one temperature to another. The transition probability is the probability of moving from one state to another. This probability depends on the current temperature, the order of generating the candidate moves, and the acceptance probability function. Markov Chain Monte Carlo (MCMC) method is applied to make a transition from one state to another (Liesenfeld and Richard 2008). The MCMC creates a chain of possible antecedents and accepts or rejects them using Metropolis algorithm.

The MCMC technique is a computational procedure for simulating a chain of states through a random walk process and consists of a Markov process and a Monte Carlo simulation (Jing and Vadakkepat 2009; Lombardi 2007). Suppose we consider a system whose evolution is described by a stochastic process consisting of random variables $\{x_1, x_2, x_3, \ldots, x_i\}$. A random variable x_i occupies a state x at discrete time i. The list of all possible states that all random variables can possibly occupy is called a state space. If the probability that the system is in state x_i+1 at time $i+1$ depends completely on the fact that it was in state x_i at time i, then the random variables $\{x_1, x_2, x_3, \ldots, x_i\}$ form a Markov chain. In the Markov Chain Monte Carlo, the transition between states is attained by adding a random noise (ε) to the current state as follows:

$$x_{i+1} = x_i + \varepsilon \tag{5.10}$$

When the current state has been attained, it is either accepted or rejected. In this chapter the acceptance of a state is decided using the Metropolis algorithm

(Metropolis et al. 1953) which has been applied extensively to solve problems of statistical mechanics. In the Metropolis algorithm, on sampling a stochastic process $\{x_1, x_2, x_3, ..., x_j\}$ consisting of random variables, random changes to x are considered and are either accepted or rejected according to the following criterion:

$$if \; E_{new} < E_{old} \; accept \; state \; (s_{new})$$
$$else$$
$$accept \; (s_{new}) \; with \; probability$$
$$\exp\{-(E_{new} - E_{old})\}$$

(5.11)

A *Cooling scheduling* is the process through which the temperature T should be reduced during simulated annealing (De Vicente et al. 2003). Experience from the physical simulated annealing dictate that the cooling rate should be sufficiently low for the probability distribution of the present state to be close to the thermodynamic equilibrium at all times during the simulation (Das and Chakrabarti 2005). The time it takes during the simulation for the equilibrium to be restored (which is also called the *relaxation time*) after a change in temperature depends on the shape of the objective function, the present temperature and the candidate generator. The ideal cooling rate is empirically obtained for each problem. The type of simulated annealing which is called *thermodynamic simulated annealing* attempts to sidestep this problem by eliminating the cooling schedule and adjusting the temperature at each step in the simulation, based on the difference in energy between the two states, in accordance to the laws of thermodynamics (Weinberger 1990). The implementation of simulated annealing is indicated in Fig. 5.1. The following cooling model is applied (Salazar and Toral 2006):

$$T(i) = \frac{T(i-1)}{1+\sigma}$$

(5.12)

Here $T(i)$ is the current temperature; $T(i-1)$ is the previous temperature and σ is the cooling rate. It must be noted that the precision of the numbers applied in the implementation of simulated annealing can have a significant effect on the outcome. A method to improve the computational time of simulated annealing is to implement either very fast simulated re-annealing or adaptive simulated annealing (Salazar and Toral 2006).

This process is repeated such that the sampling statistics for the current temperature is adequate. Then the temperature is decreased, and the process is repeated until a frozen state is achieved where $T=0$. Simulated annealing was first applied to optimization problems by Kirkpatrick et al. (1983).

5.6 Investigation and Results

This chapter implements neuro-fuzzy model and simulated annealing to build a rational counterfactual machine which is described in Fig. 5.3. This model is used in the problem of militarized interstate dispute (MID). As described by Marwala and Lagazio (2011), we use four variables associated with realist analysis and three "Kantian" variables. The first variable is *Allies*, a binary measure coded 1 if the members of a dyad (pairs of countries) are linked by any form of military alliance, and 0 in the absence of military alliance. *Contingency* is also binary, coded 1 if both states share a common boundary and 0 if they do not, and *Distance* is the logarithm, to the base 10, of the distance in kilometers between the two states' capitals. *Major Power* is a binary variable, coded 1 if either or both states in the dyad is a major power and 0 if neither are super powers. *Capability* is the logarithm, to the base 10, of the ratio of the total population plus the number of people in urban areas plus industrial energy consumption plus iron and steel production plus the number of military personnel in active duty plus military expenditure in dollars in the last 5 years measured on stronger country to weak country. The variable *Democracy* is measured on a scale where the value of 10 is an extreme democracy and a value of − 10 is an extreme autocracy and taking the lowest value of the two countries. The variable *Dependency* is measured as the sum of the countries import and export with its partner divided by the Gross Domestic Product of the stronger country. It is a continuous variable measuring the level of economic interdependence (dyadic trade as a portion of a state's gross domestic product) of the less economically dependent state in the dyad. These measures were derived from conceptualizations and measurements conducted by the Correlates of War (COW) project (Anonymous 2010).

Politically relevant population which are all dyads containing a major power are selected for the reason that it sets a hard test for prediction. Neglecting all distant dyads composed of weak states means that we ignore much of the influence with variables that are not very responsive to policy intervention (distance and national power). This exclusion makes the task difficult by decreasing the extrapolative power of such variables. Using the training and validation sampling procedure it is revealed that a strong performance is realized even when the analysis is circumscribed to the politically relevant group. By concentrating only on dyads that either encompass major powers or are contiguous, the discriminative power of the neuro-fuzzy model is tested on a difficult set of cases. The neuro-fuzzy model is trained with only highly informative data because every dyad can be considered to be at risk of experiencing a dispute, hitherto it is harder for the neuro-fuzzy model to discriminate between the two classes (dyad-years with disputes and those without disputes) for the reason that the politically relevant group is more homogeneous (e.g., closer, more inter-dependent) than the all-dyad data set.

As in Marwala and Lagazio (2011) the training data set consists of 500 conflict- and 500 non-conflict cases, and the test data consists of 392 conflict data and

392 peace data. A balanced training set, with a randomly selected equal number of conflict- and non-conflict cases was chosen to yield robust classification and stronger comprehensions on the explanation of conflicts. The data were normalized to fall between 0 and 1. The antecedent variables were *Distance, Contiguity, Major Power, Allies, Democracy, Economic Interdependency*, and *Capability* and the consequent was either peace or war. In this regard and due to normalization two countries with the largest distance between their capitals will be assigned a value of Distance of 1 while the two countries with the shortest distance between their capitals will be assigned a Distance of 0. If both countries are superpower then they will be assigned variable Major Power of 1 while if none are a major power of 0. If two countries are not allies there are assigned a value of 0 while if they are allies a value of 1. If the two countries share a border they will be assigned a Contiguity value of 1 while if they do not share a border a contiguity of 0. If the two countries have no economic interdependency the variable economic interdependency is 0 while if they have maximum economic interdependency recorded they are assigned a value of 1. For the maximum military capability the value is 1 while minimum is 0. The results obtained for modelling the relationship between the antecedent and consequent demonstrate that detection rate of conflict was 77 % and for peace was 73 % whereas the overall the detection rate of a correct outcome was 75 %.

Takagi-Sugeno neuro-fuzzy systems and simulated annealing were implemented to model militarized interstate dispute data. When these data were used in the modelling process a factual below is obtained:

If it is the case that (D=0, C=1, MJ=0.4, A=0.1, D=0.3, EI=0.1, Cap=0.6) then it will be the case that Consequent=War.

Simulated annealing is used in a manner described in Fig. 5.3 to identify the antecedent that would turn this factual into a counterfactual. In this regard the following rational counterfactual is identified that achieves peaceful outcome:

If it were the case that (D=0.7, C=1, MJ=0.4, A=.8, D=0.3, EI=0.1, Cap=0.7) then it will be the case that Consequent=Peace.

This counterfactual is deemed a rational counterfactual because it is formulated by identifying the antecedent which maximizes the attainment of a particular desired consequent and in this chapter peace.

Conclusions

This chapter introduced rational counterfactual which are counterfactuals that maximize the attainment of the desired consequent. The theory of rational counterfactuals was applied to identify the antecedent that gave the desired consequent. The results obtained demonstrated the viability of a method of identifying rational counterfactuals.

References

Akhmatskaya E, Bou-Rabee N, Reich S (2009) A comparison of generalized hybrid Monte Carlo methods with and without momentum flip. J Comput Phys 228:2256–2265

Anonymous (2010) Correlates of War Project. http://www.correlatesofwar.org/. Accessed 20 Sept 2010

Araujo E (2008) Improved Takagi-Sugeno fuzzy approach. Proceedings of the IEEE international conference on fuzzy systems, pp 1154–1158

Ata R, Kocyigit Y (2010) An adaptive neuro-fuzzy inference system approach for prediction of tip speed ratio in wind turbines. Expert Syst with Appl 37:5454–5460

Babuska R (1991) Fuzzy Modeling and Identification. PhD Thesis, Technical University of Delft

Babuska R, Verbruggen H (2003) Neuro-fuzzy methods for nonlinear system identification. Annual Rev Contr 27:73–85

Biacino L, Gerla G (2002) Fuzzy logic, continuity and effectiveness. Archive Math Logic 41:643–667

Bih J (2006) Paradigm shift—an introduction to fuzzy logic. IEEE Potentials 25(1):6–21

Birke D, Butter M, Koppe T (eds) (2011) Counterfactual thinking—counterfactual writing. de Gruyter, Berlin

Bishop CM (1995) Neural networks for pattern recognition. Oxford University, Oxford

Bryan K, Cunningham P, Bolshkova N (2006) Application of simulated annealing to the biclustering of gene expression data. IEEE Trans Inf Technol Biomed 10(3):519–525

Cano-Izquierdo J, Almonacid M, Ibarrola JJ (2010) Applying neuro-fuzzy model dfasart in control systems. Eng Appl Art Intelli 23:1053–1063

Cantor G (1874) Über eine Eigenschaft des Inbegriffes aller reellen algebraischen Zahlen. Crelles J Math 77:258–262

Celuch K, Saxby C (2013) Counterfactual thinking and ethical decision making: a new approach to an old problem for marketing education. J Mark Educ 35(2):155–167

Cox E (1994) The fuzzy systems handbook: a practitioner's guide to building, using, maintaining fuzzy systems. AP Professional, Boston

Dafflon B, Irving J, Holliger K (2009) Simulated-annealing-based conditional simulation for the local-scale characterization of heterogeneous aquifers. J Appl Geophys 68:60–70

Daftary-Kapur T, Berry M (2010) The effects of outcome severity, damage amounts and counterfactual thinking on juror punitive damage award decision making. Am J Forensic Psychol 28(1):21–45

Das A, Chakrabarti BK (2005) Quantum annealing and related optimization methods. Lect notes in Phys 679. Springer, Heidelberg

De Vicente J, Lanchares J, Hermida R (2003) Placement by thermodynamic simulated annealing. Phys Lett A 317:415–423

Devlin K (1993) The joy of sets. Springer, Berlin

Ferreirós J (1999) Labyrinth of thought: a history of set theory and its role in modern mathematics. Birkhäuser, Basel

Fogel SO, Berry T (2010) The disposition effect and individual investor decisions: the roles of regret and counterfactual alternatives. Handbook of Behavioral Finance, pp 65–80

Hájek P (1995) Fuzzy logic and arithmetical hierarchy. Fuzzy Sets and Syst 3:359–363

Hájek P (1998) Metamathematics of fuzzy logic. Kluwer, Dordrecht

Halpern JY (2003) Reasoning about uncertainty. MIT, Cambridge

He R, Hwang S (2006) Damage detection by an adaptive real-parameter simulated annealing genetic algorithm. Comput Struct 84:2231–2243

Hegel G, W F (1874) The logic encyclopaedia of the philosophical sciences, 2nd edn. Oxford University Press, London

Howlett JR, Paulus MP (2013) Decision-making dysfunctions of counterfactuals in depression: who might I have been? Front Psychiatry 4:143

Hsu Y-C, Lin S-F (2009) Reinforcement group cooperation-based symbiotic evolution for recurrent wavelet-based neuro-fuzzy systems. J Neurocomput 72:2418–2432

Hume D (1748) An enquiry concerning human understanding. Harvard Classics, vol. 37, part 3, Copyright 1910 P.F, Collier & Son

Iplikci S (2010) Support vector machines based neuro-fuzzy control of nonlinear systems. J Neurocomput 73:2097–2107

Jang J, S R (1993) ANFIS: adaptive-network-based fuzzy inference system. IEEE Trans on Syst, Man and Cybern 23:665–685

Jang JSR, Sun CT, Mizutani E (1997) Neuro-fuzzy and soft computing: a computational approach to learning and machine intelligence. Prentice Hall, Toronto

Jing L, Vadakkepat P (2009) Interacting MCMC particle filter for tracking maneuvering target. Digit signal process. doi: 10.1016/j.dsp.2009.08.011

Johansson M, Broström L (2011) Counterfactual reasoning in surrogate decision making—another look. Bioethics 25(5):244–249

Johnson P (1972) A history of set theory. Prindle, Weber & Schmidt, Boston

Kahneman D, Miller D (1986) Norm theory: comparing reality to its alternatives. Psychol Rev 93(2):136–153

Khajeh A, Modarress H (2010) Prediction of solubility of gases in polystyrene by adaptive neuro-fuzzy inference system and radial basis function neural network. Expert Syst with Appl 37:3070–3074

Kirkpatrick S, Gelatt CD, Vecchi MP (1983) Optimization by simulated annealing. Sci New Ser 220:671–680

Klir GJ, Folger TA (1988) Fuzzy sets, uncertainty, and information. Prentice Hall, New Jersey

Klir GJ, Yuan B (1995) Fuzzy sets and fuzzy logic: theory and applications. Prentice Hall, New Jersey

Klir GJ, St Clair UH, Yuan B (1997) Fuzzy set theory: foundations and applications. Prentice Hall, New Jersey

Kosko B (1993) Fuzzy thinking: the new science of fuzzy logic. Hyperion, New York

Kosko B, Isaka S (1993) Fuzzy logic. Sci Amer 269:76–81

Leach JK, Patall EA (2013) Maximizing and counterfactual thinking in academic major decision making. J Career Assess 21(3):414–429

Lewis D (1973) Counterfactuals. Blackwell, Oxford

Lewis D (1979) Counterfactual dependence and time's arrow. Noûs 13:455–476 (Reprinted in his (1986a)

Liesenfeld R, Richard J (2008) Improving MCMC, using efficient importance sampling. Comput Statistics and Data Anal 53:272–288

Lombardi MJ (2007) Bayesian inference for [Alpha]-stable distributions: a random walk MCMC approach. Comput Statistics and Data Anal 51:2688–2700

Mamdani EH (1974) Application of fuzzy algorithms for the control of a dynamic plant. Proc IEE 121:1585–1588

Marwala T (2009) Computational intelligence for missing data imputation, estimation and management: knowledge optimization techniques. IGI Global Publications, New York

Marwala T (2010) Finite element model updating using computational intelligence techniques. Springer, Heidelberg

Marwala T, Lagazio M (2004) Modelling and controlling interstate conflict. Proceedings of the IEEE international joint conference on neural networks, pp 1233–1238

Marwala T, Lagazio M (2011) Militarized conflict modeling using computational intelligence. Springer-Verlag, London

Marx K (1873) Afterword to the Second German Edition. Capital Volume 1. In Collected works, vol. 35, pp. 12–20

Mashrei MA, Abdulrazzaq N, Abdalla TY, Rahman MS (2010) Neural networks model and adaptive neuro-fuzzy inference system for predicting the moment capacity of ferrocement members. Eng Struct 32:1723–1734

Mathe P, Novak E (2007) Simple Monte Carlo and the metropolis algorithm. J Complex 23: 673–696

Metropolis N, Rosenbluth AW, Rosenbluth MN, Teller AH, Teller E (1953) Equations of state calculations by fast computing machines. J Chem Phys 21:1087

Mill JS (1843) A system of logic, The collected works of John Stuart Mill, vol. 7. A system of logic ratiocinative and inductive: being a connected view of the principles of evidence and the methods of scientific investigation, part 1, books 1–3, University of Toronto Press

Moita JMS, Correia VMF, Martins PG, Soares CMM, Soares CAM (2006) Optimal design in vibration control of adaptive structures using a simulated annealing algorithm. Compos Struct 75:79–87

Montazer GA, Saremi HQ, Khatibi V (2010) A neuro-fuzzy inference engine for farsi numeral characters recognition. Expert Syst with Appl 37:6327–6337

Novák V (1989) Fuzzy sets and their applications. Adam Hilger, Bristol

Novák V (2005) On fuzzy type theory. Fuzzy Sets and Syst 149:235–273

Novák V, Perfilieva I, Močkoř J (1999) Mathematical principles of fuzzy logic. Kluwer Academic, Dordrecht

Paya-Zaforteza I, Yepes V, Hospitaler A, Gonzalez-Vidosa F (2009) CO2-Optimization of reinforced concrete frames by simulated annealing. Eng Struct 31:1501–1508

Ratick S, Schwarz G (2009) Monte Carlo simulation. In: Kitchin R, Thrift N (eds) International encyclopedia of human geography. Elsevier, Oxford

Roese N (1997) Counterfactual thinking. Psychol Bull 121(1):133–148

Salazar R, Toral R (2006) Simulated annealing using hybrid monte carlo. arXiv:cond-mat/9706051

Sentes M, Babuska R, Kaymak U, van Nauta Lemke H (1998) Similarity measures in fuzzy rule base simplification. IEEE Trans on Syst, Man and Cybern-Part B: Cybern 28:376–386

Shaffer MJ (2009) Decision theory, intelligent planning and counterfactuals. Mind Mach 19(1): 61–92

Simioni S, Schluep M, Bault N, Coricelli G, Kleeberg J, du Pasquier RA, Gschwind M, Vuilleumier P, Annoni J-M (2012) Multiple sclerosis decreases explicit counterfactual processing and risk taking in decision making. PLoS ONE 7(12):e50718

Sugeno M (1985) Industrial applications of fuzzy control. Elsevier Science Publication Company, Amsterdam

Sugeno M, Kang G (1988) Structure identification of fuzzy model. Fuzzy Sets and Syst 28:15–33

Takagi T, Sugeno M (1985) Fuzzy identification of systems and its applications to modeling and control. IEEE Trans on Syst, Man, and Cybern 15:116–132

Talei A, Hock L, Chua C, Quek C (2010) A novel application of a neuro-fuzzy computational technique in event-based rainfall-runoff modeling. Expert Syst with Appl 37:7456–7468

Tettey T, Marwala T (2007) Conflict Modelling and Knowledge Extraction Using Computational Intelligence Methods. In: Proceedings of the 11th IEEE international conference on intelligent engineering systems, pp 161–166

von Altrock C (1995) Fuzzy logic and neurofuzzy applications explained. Prentice Hall, New Jersey

Weinberger E (1990) Correlated and uncorrelated fitness landscapes and how to tell the difference. Biol Cybern 63:325–336

Weizhong AN, Fengjuan YU, Dong F, Yangdong HU (2008) Simulated annealing approach to the optimal synthesis of distillation column with intermediate heat exchangers. Chin J Chem Eng 16:30–35

Woodward J (2003) Making things happen: a theory of causal explanation. Oxford University, Oxford

Woodward J, Hitchcock C (2003) Explanatory generalizations. Part I: a counterfactual account. Noûs 37:1–24

Wright S, Marwala T (2006) Artificial intelligence techniques for steam generator modelling. arXiv:0811.1711

Zadeh LA (1965) Fuzzy sets. Info and Control 8:338–353

Zemankova-Leech M (1983) Fuzzy relational data bases. PhD Dissertation, Florida State University

Zimmermann H (2001) Fuzzy set theory and its applications. Kluwer Academic Publishers, Boston

Chapter 6
Flexibly-Bounded Rationality in Interstate Conflict

6.1 Introduction

This chapter deals with the complicated matter of decision making which has been attempting to evolve from superstition to rationality for a long time (Vyse 2000; Foster and Kokko 2009; Marwala 2014). In particular this chapter applies the theory of flexibly-bounded rationality to the interstate conflict (Marwala 2013). There are two components of a rational decision making process and these are information and logic. The flexibly-bounded rationality theory is an extension of Herbert Simon's theory of bounded rationality which is a theory that states that on decision making, rationality is limited by the amount of information available and accessible, the cognitive limitations of the minds of the decision maker and the amount of time available to make the decisions (Simon 1957, 1991; Tisdell 1996; Stanciu-Viziteu 2012; Yao and Li 2013). The theory of flexibly-bounded rationality is a theory that states that because of recent advances in artificial intelligence the incomplete information in Simon's bounded rationality can be further completed using missing data estimation methods, the cognitive limitations of the mind on decision making can be further improved using artificial intelligence machine for decision making and the finite time of making decisions can be accelerated given the advances in computing power as advocated by Moore's Law (Liddle 2006; Moore 1965). Because of this reason, the bounds in Simon's bounded rationality are in fact flexible depending on how much artificial intelligence and computing power you put into a decision making process.

In this chapter we apply the theory of flexibly bounded rationality for decision making in militarized interstate disputes. As described by Marwala (2013), a decision making system has two elements and these are a system consisting of the statistical analysis device to process the data needed for decision making, a correlation machine which estimates missing information and a causal machine to relate the information to a decision. In this chapter, the correlation machine proposed is made out of the autoassociative machine based on the multi-layer perceptron network and particle swarm optimization method and is used to estimate missing data while the causal machine also made out of the multi-layer perceptron is used to build a

T. Marwala, *Artificial Intelligence Techniques for Rational Decision Making*,
Advanced Information and Knowledge Processing,
DOI 10.1007/978-3-319-11424-8_6

decision machine based on causal relations. Thus the causal and correlation machines are used to implement the flexibly-bounded rational decision maker.

6.2   Rational Decision Making

Rational decision making is defined as a process of attaining decisions through logic and reason in order to arrive at a point of execution of a decision (Nozick 1993; Spohn 2002; Marwala 2014). In order to understand rational decision making it is important to understand the concept of rationality. Rationality in this chapter implies logic (based on scientific principles), relevant information (it should be based on relevant information) and optimization (it should be optimized) and thus not waste energy (DeGroot 1970; Berger 1980). A decision is thus deemed to be rational and optimal if it maximizes the good that results from its consequences and at the same time minimizes the bad that is also derived from its consequences and this we define as a net utility (Marwala 2014).

Consequently a rational decision making process is a procedure by which a decision is reached using gathered relevant information and intelligence to make an optimized decision. It is considered to be rational for the reason that it uses evidence in the form of relevant information and sound and justifiable logic. Therefore, making decisions in the absence of relevant information is irrational.

Using optimization approach to describe rational decision making, we can view rational decision making as a process of identifying decision actions $d_1, d_2, ..., d_m$ given the observed information $x_1, x_2, ..., x_n$ that maximizes the expected (E) utility function U given the constraints $C_1, C_2, ..., C_n$. This can be expressed mathematically as follows:

Maximize $E\left[U(x_1, x_2, ..., x_n, d_1, d_2, ..., d_m)\right]$ (6.1)
Subject to

$$C_1(x_1, x_2, ..., x_n, d_1, d_2, ..., d_m), ..., C_m(x_1, x_2, ..., x_n, d_1, d_2, ..., d_m)   \quad (6.2)$$

In order to make a rational decision then Eq. 6.1 should be globally optimized subject to the constraints in Eq. 6.2. If a global optimum point is not achieved then the solution is not rational because it wastes resources. Furthermore, Eq. 6.1 has information $x_1, x_2, ..., x_n$ and this information should all be identifiable otherwise the optimization problem will be truncated a situation that Herbert Simon deems bounded rational problem which will be discussed in detail later. Solving Eq. 6.1 and 6.2 should be conducted efficiently.

6.2.1   Inference and Rational Decision Making

According to the dictionary inference is defined as "a conclusion reached on the basis of evidence and reasoning". There are many techniques that have been proposed to enable inference and these include artificial intelligence (MacKay 2003; Russell and Norvig 2003), inductive inference (Jeffrey 1979; Angluin and Smith 1983) and adductive inference (O'Rourke and Josephson 1997; Psillos 2009). In earlier chapter we stated that a function can either be a causal or a correlation function and can be written as follows:

$$y = f(x) \qquad (6.3)$$

Here f is the functional mapping and for a causal function there is a flow of information from x to y and if it is correlation function then there is a relationship between x and y but no flow of information between the two variables. This idea of inferring y from x is either a correlation or a causal exercise. In order to infer y all one needs is x (evidence) and f (the logic that describes the data). Now if y is some characteristics in the future of x this relationship basically says that in order to infer the future characteristics defined by variable y then all one needs to do is to rearrange the present (x and f). Therefore, rational future prediction is nothing but a rearrangement of the present and therefore *it is impossible to rationally predict what does not already exists.*

Theory of Rational Expectations

The theory of rational expectation is a theory that states that agents do not make systematic mistakes when making predictions for the future and therefore the error they make is random. This then implies that if one averages the error of agents when predicting the future then the average error will be on average zero. From this theory came the economic theory of the efficient market hypotheses which states that markets are efficient and that they reflect all the available information in the market This theory of course assumes that the agents do not have an intention to act irrationally. For example, suppose John Shavhani is a President of a country called Duthuni. He decides to siphon money to his own private offshore account, a process known as corruption. John is not acting rationally because this will destroy his country and the well-being of the citizens including his own family. However, the performance of his country will be biased towards the worst because of his deliberate corrupt activity and the theory of rational expectation cannot be accurately applied to analyse the future economic performance of the country Duthuni because of the bias introduced by President Shahani in his illicit activities.

The theory of modelling rational expectations was first introduced by Muth (1961) within the context of modelling economic agents (Sargent 1987). Applications of the theory of rational expectations in modelling economic agents are found in Savin (1987), Snowdon et al. (1994) and Janssen (1993). There has been extensive criticism of the theory of rational expectations some of which go as far as proclaiming this theory to be fundamentally irrational (Lodhia 2005).

6.2.2 Theory of Rational Choice

The rational choice theory states that individual choose the cause of actions that maximize utility (benefit less cost) (Anand 1993; Blume and Easley 2008; Sen 2008; Grüne-Yanoff 2012). The philosophical roots of the idea that the best policy is that which maximizes utility is known as utilitarianism and was studied by theorists such as Jeremy Bentham and John Stuart Mill (Adams 1976; Anscombe 1958; Bentham 2009; Mill 2011). It assumes that (Bicchieri 1993): (1) it is possible to rank all decision actions in order of preference (completeness); (2) it is possible to compare all decision actions; and (3) If decision action A is better that B out of the decision choice set $\{A, B\}$, then presenting a new decision action Y to create a decision choice set $\{A, B, Y\}$ does not make B preferable to A. Formulated differently, the theory of rational choice states that an optimal decision is constructed based on the product of the effect of the decision and its probability of incidence (Allingham 2002). Nevertheless, this theory was observed to be insufficient for the reason that it does not take into consideration where the person will be compared to where they originally were before making a decision. Kahneman and Tversky (1979) improved the theory of rational choice by introducing the prospect theory which includes the reference situation to assess the decision (Kahneman 2011; Marwala 2014).

6.2.3 Theory of Rational Counterfactuals

As described in the previous chapter, counterfactual can be comprehended by separating these two words into two segments *counter* and *factual*. Factual is an event that has happened for example: Saddam Hussein invaded Kuwait and consequently George Bush declared war on Iraq. *Counter* means the opposite and in the light of the factual above: If Saddam Hussein did not invade Kuwait and consequently George Bush would not have declared war on Iraq. Of course a counterfactual can be an imaginary concept and, therefore, the fundamental question that needs to be asked is: How do we know what would have happened if something did not happen? It is possible to estimate what might have happened and a theory of rational counterfactuals as described in the previous chapter is able to achieve this task. Rational counterfactuals are those counterfactuals in the counterfactual universe corresponding to a given factual that maximize a particular goal. If we have the factual and the counterfactual then the difference between the two can be used to establish whether there is a causal relationship between the independent variable and the dependent variable.

6.2.4 Bounded-Rational Decision Making

In rational decision making all information are used to make an optimized decision in the shortest time. This is not practically possible for the reason that information necessary to make decisions is very often not available and the limitation of the infrastructure to make sense of the information and the limited time needed to process such information (Simon 1991). This is what is called the theory of bounded rationality and was proposed by Nobel Laureate Herbert Simon (Simon 1957; Tisdell 1996; Murata et al. 2012). The theory of bounded rationality has not replaced the theory of rationality which was described in earlier section but places limitations of incomplete information, imperfect decision making and limited time for completing the decision making process, on the applicability of the theory of rationality. Herbert Simon coined a term *satisficing,* thereby hybridizing the terms satisfying and sufficing, which is the concept of making optimized decision under the limitations that the data used in making such decisions are imperfect and incomplete while the model used to make such decisions is inconsistent and imperfect. The consequence of this theory on many economic systems is pretty considerable (Simon 1991; Aviad and Roy 2012). It interrogates the reliability of the theory of the efficient market hypothesis particularly given the fact that players in the market are at best only rational within the bounds of limited information and limited model to make sense of the information. Bounded rationality has been studied by many researchers such as Gama (2013) in stream mining, Jin et al. (2013) to build a computer virus propagation model as well as Jiang et al. (2013) in water saving and pollution prevention. In bounded rationality Eq. 6.1 and 6.2 can be modified to become:

$$\text{Maximize } E\big[U(x_1,x_2,\ldots,x,d_1,d_2,\ldots,d_m)\big] \tag{6.4}$$
Subject to

$$C_1(x_1,x_2,\ldots,x_p,d_1,d_2,\ldots,d_m),\ldots,C_p(x_1,x_2,\ldots,x_p,d_1,d_2,\ldots,d_m) \tag{6.5}$$

Here the information is truncated $(x_p+1, x_{p+}2,\ldots x_n)$ because some of the information are missing. Equation 6.3 and 6.4 also depends on the amount of computational capabilities that is put into the decision making process and as we know this is variable due to Moore's Law. Now that we have described the theory of bounded rationality, which resulted in the limitations of the applicability of the theory of rational decision making, the next section describes the theory of flexibly-bounded rationality which is theory applied in this chapter.

6.3   Flexibly-Bounded Rational Decision Making

This section discusses the theory of flexibly-bounded rationality and this theory is based on the following assumptions (Marwala 2014):

1. Rational decision making is a process of making optimized decisions based on logic.
2. Rationality whether as in a decision process or decision action is indivisible. In other words, you cannot be half rational and half irrational. If you are half rational and half irrational you are being irrational.
3. The principle of bounded rationality does not truncate the theory of rationality but merely specifies the bounds within which the principle of rationality is applied.

The theory of flexibly bounded rationality is a characterization of the properties of the bounds that define the theory of bounded rationality. This is due to the fact that there have been advances in information processing techniques, missing data estimation technology, computer processing capability due to Moore's Law and artificial intelligence methods, which make the bounds in the theory of bounded rationality flexibility. Tsang (2008) proposed that computational intelligence determines effective rationality. What Tsang implies is that there is a degree of rationality, a situation which is not true. Rationality cannot be divided, it is either there or absent. The model of flexibly-bounded rationality can thus be expressed mathematically as follows:

$$\text{Maximize } E\left[U(x_1, x_2, \ldots, x'_{p+1}, \ldots x'_n, d_1, d_2, \ldots, d_m)\right] \tag{6.6}$$
Subject to

$$C_1(x_1, x_2, \ldots, x'_{p+1}, \ldots x'_n, d_1, d_2, \ldots, d_m), \ldots, C_m(x_1, x_2, \ldots, x'_{p+1}, \ldots x'_n, d_1, d_2, \ldots, d_m) \tag{6.7}$$

In these equations missing information are estimated $x'_{p+1}, \ldots x'_n$ and these are used to estimate the expected utility (U). Estimating the expected utility is done using artificial intelligence method in a computational device whose capability is increasing as specified by Moore's Law instead of the human brain and this move the bounds of rationality. The theory of flexibly-bounded rationality only expands the bounds within which the principle of rationality is applied and that the theory of bounded rationality only prescribes the bounds within which the principle of rationality is applied. The next section introduces two elements that flex the bounds of bounded rationality theory and these are missing data estimation framework and artificial intelligence. The flexibly bounded rationality method is illustrated in Fig. 6.1. In this figure the missing data is estimated and the rational decision that maximizes utility is estimated using artificial intelligence method. The next section describes missing data estimation method.

6.4 Missing Data Estimation

To flex the bounds of the theory bounded rationality we estimate missing data to reduce information that is not observed or known. Missing data has been studied extensively and a comprehensive review on this subject can be found in the seminal book by Marwala (2009). There have been numerous approaches that have been suggested to estimate missing data and these include autoassociative learning machine to capture variables' interrelationships and optimization method to identify missing values (Abdella and Marwala 2006; Nelwamondo and Marwala 2007; Pantanowitz and Marwala 2009; Hlalele et al. 2009). In this chapter missing data is estimated using autoassociative Bayesian multi-layer perceptron neural networks to capture the model that describes the data and particle swarm optimization to estimate the missing values (Marwala 2009; Mthembu et al. 2010). The autoassociative Bayesian neural network missing data estimation technique used in this chapter creates a model which reproduces its input as the output. From this model as it was done in Marwala (2009) a missing data estimation equation can be written as follows:

Fig. 6.1 Flexibly bounded
rationality model

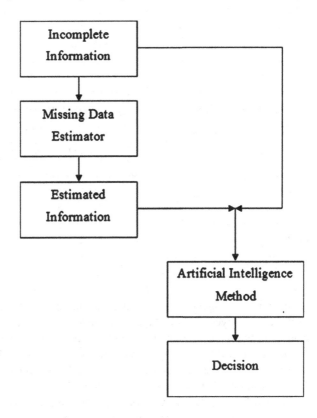

$$E = \left\| \left(\left\{ \begin{matrix} x_k \\ x_u \end{matrix} \right\} - f \left(\left\{ \begin{matrix} x_k \\ x_u \end{matrix} \right\} \{w\} \right) \right) \right\| \tag{6.8}$$

Here E is the fitness function, f is the autoassociative model, $\{w\}$ is the network
weight, $\| \; \|$ is the Euclidean norm, x_k is the known data matrix and x_u is the un-
known data matrix and these are matrices because we are doing multiple missing
data estimation. Here to estimate the missing input values, Eq. 6.8 is minimized and
in this chapter this is achieved using particle swarm optimization while f is estimated using a Bayesian multi-layer perceptron neural networks and these techniques
are discussed in detail later in the chapter. The missing data estimation framework
discussed in this section is shown in Fig. 6.2.

6.5   Intelligent Machines: Bayesian Multi-Layer
Perceptron Network

The core of the theory of flexing the bounds in the theory of bounded rationality is
the use of an artificial intelligent machine for decision making process because it
is deemed to be more consistent than the human brain. There are numerous types

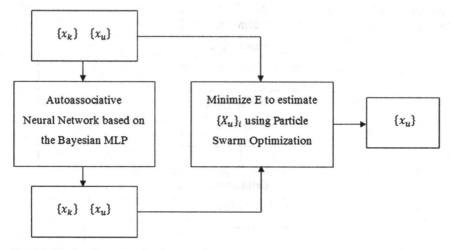

Fig. 6.2 Missing data estimation framework

of artificial intelligence methods that can be applied to make decisions and one of these is a Bayesian multi-layer perceptron neural network which is used in this chapter to build an associative network in Fig. 6.2 and an artificial intelligence decision maker in Fig. 6.1. Bayesian multi-layer perceptron (MLP) neural network is an artificial intelligence method which is used to learn patterns from data (Lagazio and Marwala 2005; Patel and Marwala 2006; Marwala 2006; Mistry et al. 2006).

The MLP is a network which relates the input and the hidden layer as well as between the hidden layer and the output using adjustable weight parameters. The MLP model is able to estimate any relationship between the input and the output provided the number of hidden neurons is sufficiently large. In the MLP network there are interconnected cross-couplings between the cause and the hidden nodes, and between the hidden nodes and the effect. Given the input variable x and the output y, a mapping function between the input and the output may be expressed as follows (Bishop 1995):

$$y = f_{\text{output}} \left(\sum_{j=1}^{M} w_j f_{\text{hidden}} \left(\sum_{i=0}^{N} w_{ij} x_i \right) + w_0 \right) \qquad (6.9)$$

here N is the number of inputs units, M is the number of hidden neurons, x_i is the ith input unit, w_{ij} is the weight parameter between input i and hidden neuron j and w_j is the weight parameter between hidden neuron j and the output neuron. The activation function $f_{\text{hidden}}(*)$ is the hyperbolic tangent function whereas $f_{\text{output}}(*)$ is logistic for the artificial intelligence decision maker in Fig. 6.1 and linear for the missing data estimation method in Fig. 6.2.

There are two processes that can be used to train an MLP network and these are the maximum likelihood and the Bayesian procedures. Maximum likelihood technique finds network weights that maximize the capacity of a trained MLP to better forecast the observed data whereas the Bayesian method creates the probability distribution of the network model given the measured data.

In this chapter we estimate the weights in Eq. 6.9 using the Bayesian approach which uses the Bayes' theorem (MacKay 1991; MacKay 1992; Bishop 1995):

$$P(w\,|\,[D]) = \frac{P([D]\,|\,w)P(w)}{P([D])} \tag{6.10}$$

Here $P(w)$ is the probability distribution function of the weight-space in the absence of the data (prior distribution function) and $[D] \equiv (y_1, ..., y_N)$ is a matrix of the data, $P(w|[D])$ is the posterior probability distribution function, $P([D]|w)$ is the likelihood function and $P([D])$ is the evidence. According to Bishop (1995) Eq. 6.10 may written as follows for the sum-of-squares of the error function:

$$P(w\,|\,[D]) = \frac{1}{Z_s} \exp\left(\beta \sum_n^N \sum_k^K \{t_{nk} - y_{nk}\}^2 - \sum_j^W \frac{\alpha_j}{2} w_j^2 \right) \tag{6.11}$$

where t_{nk} is the target output with k being the index for the output units, n is the index for the training example, N is the number of training examples, K is the number of network outputs, W is the number of network weights, whereas α and β are hyper-parameters and

$$Z_S(\alpha, \beta) = \left(\frac{2\pi}{\beta}\right)^{N/2} + \left(\frac{2\pi}{\alpha}\right)^{W/2} \tag{6.12}$$

Equation 6.11 is solved in this chapter by sampling the posterior probability space using the hybrid Monte Carlo which can be used for this task are described in the next sections.

Hybrid Monte Carlo (HMC) is class of Monte Carlo methods which are used for sampling a probability distribution function (Widom et al. 2014; Takaishi 2014; Günay et al. 2014; Jenkins et al. 2013). Monte Carlo techniques were derived from the way casinos operate and the name was inspired by the gambling hub in Monaco. In this setting one can infer the probability of winning from actually observing gambling in a casino. In Monte Carlo simulation one can thus infer the probability of winning by simulating the gambling process. This is achieved by defining the space in which Monte Carlo sampling is conducted, randomly sampling the space, putting the random sample into a deterministic model and aggregating the results.

A type of a Monte Carlo method is the Markov chain Monte Carlo (MCMC) which is a sampling technique where a Markov chain is used to make a transition from one state to another (Kim et al. 2014; Chen et al. 2014; Ahmed 2014). To understand the MCMC technique it is necessary to understand the property of a Markov process which is that the conditional probability distribution for the system at subsequent step rests only on the present state of the system. This means a state will be accepted or not depending on the probability distribution being sampled as well as the current state relative to the previous state.

The HMC method combines the Markov Chain Monte Carlo and the gradient based optimization technique and this gradient information increases the probability of sampling through the regions of higher probabilities and increases the rate of convergence to a stationary probability distribution function. The HMC method is a Markov chain method and makes dynamic moves which apply the Hamiltonian dynamics and allows sampling states with constant total energy and stochastic moves which allows the method to sample states with different total energy.

This chapter uses the Hamiltonian mechanics in the hybrid Monte Carlo method. As is known in physics, the Hamiltonian is the sum of the kinetic and potential energy of the system. The potential energy of a system is a function of its position while the kinetic energy of the system is a function momentum which is a function of its velocity. The position of all molecules at a specific time in a physical system is known as the *state space*. The canonical distribution of the system's kinetic energy is (Neal 1992):

$$P(\{p\}) = \frac{1}{Z_K} \exp(-K(\{p\}))$$

$$= (2\pi)^{-n/2} \exp(-\frac{1}{2} \sum_i p_i^2) \tag{6.13}$$

here p_i is the momentum of the ith variable and p must not to be confused with P, which designates the probability. The parameter, p_i is an imaginary variable that puts this method in a molecular dynamics framework. The *Hamiltonian* of the system can be written as (Neal 1992):

$$H(w, p) = \beta \sum_{}^{N} \sum_{k}^{K} \{y_{nk} - t_{nk}\}^2 + \frac{\alpha}{2} \sum_{j=1}^{W} w_j^2 + \frac{1}{2} \sum_{i}^{W} p_i^2 \tag{6.14}$$

In Eq. 6.14, the first two terms are the potential energy of the system and the last term is the kinetic energy. The distribution over the phase space, *i.e.*, position and momentum, can be written as (Neal 1992):

$$P(w, p) = \frac{1}{Z} \exp(-H(w, p)) = P(w \mid D)P(p) \tag{6.15}$$

The phase space can be expressed in the Hamiltonian dynamics by using the derivative of the 'position' and 'momentum' in terms of time τ The expression 'position' applied in this chapter is the neural network weights. The dynamics of the Hamiltonian consequently can be expressed as follows (Neal 1992):

$$\frac{dw_i}{d\tau} = +\frac{\partial H}{\partial p_i} = p_i \tag{6.16}$$

$$\frac{dp_i}{d\tau} = +\frac{\partial H}{\partial w_i} = -\frac{\partial E}{\partial p_i} \tag{6.17}$$

Eq. 6.16 and 6.17 cannot be solved exactly and are discretized using a 'leapfrog' technique as follows (Neal 1992):

$$\hat{p}_i(\tau + \frac{\varepsilon}{2}) = \hat{p}_i(\tau) - \frac{\varepsilon}{2}\frac{\partial E}{\partial w_i}(\hat{w}(\tau)) \qquad (6.18)$$

$$\hat{w}_i(\tau + \varepsilon) = \hat{w}_i(\tau) + \varepsilon \hat{p}_i(\tau + \frac{\varepsilon}{2}) \qquad (6.19)$$

$$\hat{p}_i(\tau + \varepsilon) = \hat{p}_i(\tau + \frac{\varepsilon}{2}) - \frac{\varepsilon}{2}\frac{\partial E}{\partial w_i}(\hat{w}(\tau + \varepsilon)) \qquad (6.20)$$

Eq. 6.18 ensures a small half step for the momentum vector, $\{p\}$, and, Eq. 6.19, changes a full step for the 'position', $\{w\}$, and, Eq. 6.20, alters a half step for the momentum vector, $\{p\}$. These three steps combined offer a single leapfrog iteration that approximates the '*position*' and '*momentum*' of a system at time $\tau + \varepsilon$ from the '*position*' and 'momentum' at time τ. This discretization is reversible in time and it almost conserves the Hamiltonian and conserves the volume in the phase space (Neal 1992).

The HMC implementation follows a series of paths from an initial state, *i.e.*, '*positions*' and '*momentum*', and moves in some direction in the state space for a given length of time and accepts the final state using the Metropolis Algorithm. Consequently, states with high probability form a high proportion of the Markov chain and those with a low probability form a low proportion of the Markov chain.

For a stated leapfrog step size, ε_0, and the number of leapfrog steps, L, the dynamic transition of the hybrid Monte Carlo method is applied as explained by Neal (1992), and Bishop (1995):

1. Select randomly the direction of the trajectory, λ, to be either -1 for a backwards trajectory or $+1$ for forwards trajectory.
2. Begin from the first state, $(\{w\}, \{p\})$, perform L leapfrog steps with the step size $\varepsilon = \varepsilon_0(1 + 0.1k)$ leading to state $(\{w\}^*, \{p\}^*)$. Here, ε_0 is a selected fixed step size and k is a number selected from a uniform distribution and is between 0 and 1.
3. Reject or accept $(\{w\}^*, \{p\}^*)$ by using the Metropolis criterion. If the state is accepted then the new state becomes $(\{w\}^*, \{p\}^*)$. If rejected, the old state, $(\{w\}, \{p\})$, is reserved as the new state.

The Metropolis algorithm is as follows (Metropolis et al. 1953):
- if $p_{new} > p_{old}$ *accept state* (s_{new})
- else generate a random value γ uniformly distributed between 0 and 1.
- *if* $^{p_{new}}/_{p_{old}} > \gamma$ then accept s_{new}
- Otherwise reject and return to s_{old}.

Parameter L needs to be high to allow rapid search of the state space. The selections of ε_0 and L influence the time at which the simulation converges to a stationary posterior probability distribution and the correlation between the states accepted.

Systematic errors are not introduced by the leapfrog discretization because of occasional rejection of states. In the hybrid Monte Carlo method, the step size $\varepsilon = \varepsilon_0(1+0.1k)$ where k is uniformly distributed between 0 and 1 is not static and this guarantees that the step size for each trajectory is changed in order to reject states that have high correlation. This can also be achieved by changing the leapfrog steps. The application of the Bayesian technique in neural networks results in weight vectors with corresponding mean and standard deviation. For that reason, the output parameters have a probability distribution. Applying probability theory, the distribution of the output vector $\{y\}$ given input vector $\{x\}$ can be written as follows (Bishop 1995):

$$p(\{y\}|\{x\},D) = \int p(\{y\}|\{x\},\{w\})p(\{w\}|D)d\{w\} \qquad (6.21)$$

The HMC method is used to identify the distribution of the weight vectors, and consequently, of the output parameters. The integral in Eq. 6.21 may be estimated as follows (Bishop 1995; Neal 1992):

$$I \equiv \frac{1}{L}\sum_{i=1}^{L} f(\{w\}_i) \qquad (6.22)$$

Here, L is the number of reserved states and f is the MLP network. The application of a Bayesian method to the neural network results in the mapping of the weight vector between the input and output having a probability distribution.

6.6   Particle Swarm Optimization

The data estimation procedure in Fig. 6.2 involves the use of an optimization method and in this chapter particle swarm optimization (PSO) is used. PSO is a stochastic, population-based evolutionary method that is widely used to optimize functions (Marwala and Lagazio 2011; Garšva and Danenas 2014; Sadeghi et al. 2014; Liu et al. 2014; Palma et al. 2014; Valdez et al. 2014). It was inspired by the social and psychological characteristics of the intelligence of swarm. Swarm behaviours are behaviours of living organisms which behave as a group and this behaviour is observed in bees, ants, birds, fish etc. For example, a South African named Eugene Marais in his seminal work was the first person to study the behaviour of white ants the mould of which is shown in Fig. 6.3 (Marais 2009).

The reason why white ants, which are incredibly small compared to their mould and do not have eyes so they navigate by feeling, can be able to build a structure that is as complicated and vast as what is shown in Fig. 6.3 is that they work as a team and they invoke individual as well as group intelligence. This swarm intelligence (individual and group) allows for the understanding of social behavior. Every element of a swarm (white ants) behaves by balancing between its individual and group knowledge. The version of swarm intelligence that is used in this chapter is

Fig. 6.3 The mould of a
white ant

the particle swarm optimization which is inspired by the manner in which a swarm of birds find its roost and which was proposed by Kennedy and Eberhart (1995). An example of a swarm of geese is shown in Fig. 6.4.

The PSO method is applied to minimize Eq. 6.8 and thereby estimate missing data. To achieve this, a social network articulating a population of possible solutions is generated randomly and these individuals in this social network are allocated neighbors to interrelate with and these individuals are so-called particles, therefore the expression particle swarm optimization. Afterwards, a procedure to improve these particles is begun by assessing the error of each particle as indicated in Eq. 6.8. Every particle recalls the location where it had its best success as measured by the error in Eq. 6.8. The best solution of the particle is titled the *local best* and each particle shares this information on the local best to its neighbors and perceives its neighbors' success. The procedure of exploring the feasible space is directed by these successes and the population eventually converges to a minimum point.

In particle swarm optimization striking the balance between sampling for a good solution and using other particles' successes is important because if the sampling for a solution is inadequate then the PSO converges to a local optimum solution whereas if the gains of particles are not exploited then the PSO does not converge. The reason why PSO is considered a good model is because it is computationally efficient, is simple to implement, and has few adjustable parameters when compared to other competing evolutionary programming techniques such as genetic algorithm.

Fig. 6.4 Swarm of geese

Particle swarm optimization technique is applied by initializing a population of random potential solutions, each considered as particles and is assigned a random velocity and position. Each particle is denoted by two features which are the position vector of particle i at step k designated $p_i(k)$ and the respective velocity vector of particle i at step k designated by $v_i(k)$. Positions and velocities of particles are initially randomly sampled and the subsequent positions and velocities are approximated using the position of the best solution that a particular particle has come across during the simulation labelled $pbest_i$ and the best particle in the swarm, which is entitled $gbest(k)$. The next velocity of a particle i can be calculated as follows (Kennedy and Eberhart 1995; Marwala and Lagazio 2011):

$$v_i(k+1) = \gamma v_i(k) + c_1 r_1 (pbest_i - p_i(k)) + c_2 r_2 (gbest(k) - p_i(k)) \qquad (6.23)$$

Here γ is the inertia of the particle, c_1 and c_2 are the 'trust' parameters, and r_1 and r_2 are randomly generated numbers between 0 and 1.

The first expression in Eq. 6.23 is the present motion, the second expression denotes particle memory and the third term group influence. For a particle i the position p_i at step $k+1$ can be estimated from the position p_i at step k and velocity v_i at step $k+1$ as follows (Kennedy and Eberhart 1995; Marwala 2009; Marwala and Lagazio 2011):

$$p_i(k+1) = p_i(k) + v_i(k+1) \qquad (6.24)$$

The inertia of the particle controls the effect of the past velocity of the particle on the present velocity and these balances the exploratory feature of the PSO with a high inertia facilitating global exploration and a low inertia facilitating local exploration.

Parameter c_1 is the confidence the current particle has on itself and this indicates individual intelligence c_2 is the confidence the current particle has on the population and this is group intelligence. Parameters r_1 and r_2 are randomly generated numbers between 0 and 1 and they control the exploratory capacity of the PSO.

6.7   Experimental Investigations: Interstate Conflict

This chapter implements Bayesian MLP and particle swarm optimization methods in interstate conflict and demonstrate the flexing of the bounds of the theory of bounded rationality. As described by Marwala and Lagazio (2011), we use seven variables to predict interstate conflict and these variables are *Allies, Contingency, Distance, Major Power, Capability, Democracy* and *Dependency* from the Corre-lates of War (COW) project (Anonymous 2010).

The autoassociative Bayesian network was built using 500 conflict- and 500 non-conflict cases and had seven inputs, seven outputs, six hidden nodes, hyperbolic tangent function in the hidden layer and linear function in outer layer. The autoas-sociative Bayesian MLP was trained using the HMC with 200 retained states. The test data consisted of 392 conflict data and 392 peace data. A balanced training set, with a randomly selected equal number of conflict- and non-conflict cases was cho-sen to yield robust classification and stronger comprehensions on the explanation of conflicts. The first experiment considered the variable *Allies* to missing and then estimated it and then used it to predict the conflict status on the 392 conflict and 392 on-conflict data. To perform missing data estimation with variable *Allies* missing, the PSO with a population of 20 and 100 simulation iterations was implemented. Then after the missing data was estimated then a Bayesian MLP with 7 input vari-ables and a peace/conflict status as output was constructed using the hyperbolic tangent function in the hidden layer, logistic function in the outer layer and 200 retained states and this network is called the peace/conflict Bayesian MLP predictor. When there were no missing data and the peace/conflict Bayesian MLP predictor was used for prediction the accuracy obtained was 73.5 %. When the variable *Allies* was missing and then peace/conflict status was estimated using the peace/conflict Bayesian MLP predictor the accuracy of 69.7 % was obtained while the accuracy of 71.5 % was obtained when the peace/conflict Bayesian MLP predictor was used with the estimated variable *Allies*. When the variables *Allies* and *Democracy* were missing and then peace/conflict status was estimated using the peace/conflict Bayes-ian MLP predictor the accuracy of 61.4 % was obtained while the accuracy of 66.8 % was obtained when the peace/conflict Bayesian MLP predictor was used with the estimated variables *Allies* and *Democracy*. When the variables *Allies, Democracy* and *Capability* were missing and then peace/conflict status was estimated using the peace/conflict Bayesian MLP predictor the accuracy of 53.1 % was obtained while the accuracy of 52.8 % was obtained when the peace/conflict Bayesian MLP predic-tor was used with the estimated variables *Allies, Democracy* and *Capability*. The characteristics of the Bayesian MLP predictor are shown in Table 6.1 while the results obtained are shown in Table 6.2.

Table 6.1 Network characteristics of the peace/conflict Bayesian MLP predictor

Variables missing	None	Allies	Allies, democracy	Allies, democracy, capability
Number of input of the Bayesian MLP predictor	7	6	5	4
Number of input of the Bayesian MLP predictor	6	5	4	3
Number of output of the Bayesian MLP predictor	1	1	1	1

Table 6.2 Classification results

Number of missing data	0	1	2	3
Accuracy by using only available variables (%)	73.5	69.7	61.4	53.1
Accuracy by using approximated data (%)	73.5	71.5	66.8	52.8

The Bayesian MLP predictor had hyperbolic tangent function in the hidden layer and logistic function in the outer layer while all missing data estimation Bayesian MLP had hyperbolic tangent function in the hidden layer and linear function in the outer layer. From the results obtained it is quite evident that it is better to estimate missing variables and then make a decision based on information that include estimated variables than ignoring missing variables all together. Ignoring missing variables in decision making is bounded rationality decision making whereas estimating missing variables is moving the bounds in the theory of bounded rationality. In addition to moving the bounds in the bounded rational decision making the use of Bayesian MLP for decision making becomes more efficient as computational power increases in accordance to Moore's Law thereby moving the bounds again.

Conclusions

This chapter applied the theory of flexibly bounded rationality to interstate conflict. Flexibly bounded rationality is a theory that observes that given the fact that missing information can be estimated to a certain extent, decision making processing power can be increased due to Moore's Law and that decision making can be be enhanced using artificial intelligence methods, then the bounds of bounded rationality theory are flexible. This theory was successfully applied to the problem of interstate conflict.

References

Abdella M, Marwala T (2006) The use of genetic algorithms and neural networks to approximate missing data in database. Comput Inform 24:1001–1013
Adams RM (1976) Motive utilitarianism. J Philos 73(14):467–481

Ahmed EA (2014) Bayesian estimation based on progressive type-II censoring from two-parameter bathtub-shaped lifetime model: an Markov chain Monte Carlo approach. J Appl Statist 41(4):752–768

Allingham M (2002) Choice theory: a very short introduction. Oxford University Press, Oxford

Anand P (1993) Foundations of rational choice under risk. Oxford University Press, Oxford

Angluin D, Smith CH (1983) Inductive inference: theory and methods. Comput Surv 15(3):237–269

Anonymous (2010) Correlates of war project. http://www.correlatesofwar.org/. Accessed 20 Sep 2010

Anscombe GEM (1958) Modern moral philosophy. Philos 33(124):1–19

Aviad B, Roy G (2012) A decision support method, based on bounded rationality concepts, to reveal feature saliency in clustering problems. J Decis Support Syst 54(1):292–303

Bentham J (2009) An introduction to the principles of morals and legislation (Dover Philosophical Classics). Dover Publications Inc. Mineola, New York

Berger JO (1980) Statistical decision theory and Bayesian analysis. Springer, Heidelberg

Bicchieri C (1993) Rationality and coordination. Cambridge University Press, Cambridge

Bishop CM (1995) Neural networks for pattern recognition. Oxford University Press, Oxford

Blume LE, Easley D (2008) Rationality. The new palgrave dictionary of economics 2nd edn. Abstract, Palgrave Macmillan, Basingstoke

Chen J, Choi J, Weiss BA, Stapleton L (2014) An empirical evaluation of mediation effect analysis with manifest and latent variables using Markov Chain Monte Carlo and alternative estimation methods. Struct Equ Model 21(2):253–262

DeGroot M (1970) Optimal statistical decision. McGraw-Hill, New York

Foster KR, Kokko H (2009) The evolution of superstitious and superstition-like behaviour. Proc Biol Sci 276(1654):31–37

Gama J (2013) Data stream mining: the bounded rationality. Informatica 37(1):21–25

Garšva G, Danenas P (2014) Particle swarm optimization for linear support vector machines based classifier selection. Nonlinear Anal Model Control 19(1):26–42

Grüne-Yanoff T (2012) Paradoxes of rational choice theory. In: Roeser S, Hillerbrand R, Sandin P, Peterson M (eds) Handbook of risk theory. Springer Netherlands, pp 499–516, doi:10.1007/978-94-007-1433-5_19

Günay M, Şarer B, Kasap H (2014) The effect on radiation damage of structural material in a hybrid system by using a Monte Carlo radiation transport code. Ann Nucl Energy 63:157–161

Hlalele N, Nelwamondo FV, Marwala T (2009) Imputation of missing data using PCA, neurofuzzy and genetic algorithms. 15th International Conference, ICONIP 2008, part 4, Lecture notes in computer science, vol. 5507, pp 485–492

Janssen MCW (1993) Microfoundations: a critical inquiry. Routledge, London. International library of critical writings in economics, vol. 19. Aldershot, Hants, Elgar. pp 3-23. ISBN 978-1-85278-572-7

Jeffrey RC (ed) (1979) Studies in inductive logic and probability. University of California, Berkeley

Jenkins R, Curotto E, Mella M (2013) Replica exchange with smart Monte Carlo and hybrid Monte Carlo in manifolds. Chem Phys Lett 590:214–220

Jiang R, Xie J, Wang N, Li J (2013) Evolution game analysis of water saving and pollution prevention for city user groups based on bounded rationality. Shuili Fadian Xuebao/J Hydroelectr Eng 32(1):31–36

Jin C, Jin S-W, Tan H-Y (2013) Computer virus propagation model based on bounded rationality evolutionary game theory. Secur Commun Netw 6(2):210–218

Kahneman D (2011) Thinking, fast and slow. Macmillan, New York

Kahneman D, Tversky A (1979) Prospect theory: an analysis of decision under risk. Econometrica 47(2):263–291

Kennedy J, Eberhart R (1995) Particle swarm optimization. Proceedings of IEEE International Conference on Neural Networks IV. pp 1942–1948

Kim S, Kim G-H, Lee D (2014) Bayesian Markov chain Monte Carlo model for determining optimum tender price in multifamily housing projects. J Comput Civ Eng 28(3), (art. no. 06014001)

Lagazio M, Marwala T (2005) Assessing different Bayesian neural network models for militarized interstate dispute. Soc Sci Comput Rev 24(1):1–12

Liddle DE (2006) The wider impact of Moore's Law. Solid State Circuits Soc Newsl 11(5):28–30

Liu H, Ding G, Wang B (2014) Bare-bones particle swarm optimization with disruption operator. Appl Math Comput 238:106–122

Lodhia HC (2005) The irrationality of rational expectations-an exploration into economic fallacy, 1st edn. Warwick University Press, Coventry

MacKay D (1991) Bayesian methods for adaptive models. PhD thesis, California Institute of Technology

MacKay DJC (1992) Bayesian methods for adaptive models. PhD thesis, 2nd edn California University of Technology

MacKay DJC (2003) Information theory, inference, and learning algorithms. Cambridge University Press, Cambridge, ISBN 0-521-64298-64291

Marais E (2009) The soul of the white ant, 1937. First published as Die Siel van die Mier in 1925, the Philovox

Marwala T (2006) Genetic approach to Bayesian training of neural networks. Proceedings of the IEEE International Joint Conference on Neural Networks, BC, Canada, pp 7013–7017

Marwala T (2009) Computational intelligence for missing data imputation, estimation and management: knowledge optimization techniques. IGI Global, New York

Marwala T (2013) Flexibly-bounded rationality and marginalization of irrationality theories for decision making. arXiv:1306.2025

Marwala T (2014) Causality, correlation and artificial intelligence for rational decision making. World Scientific, Singapore (in press)

Marwala T, Lagazio M (2011) Militarized conflict modeling using computational intelligence techniques. Springer, Heidelberg

Metropolis N, Rosenbluth AW, Rosenbluth MN, Teller AH, Teller E (1953) Equations of state calculations by fast computing machines. J Chem Phys 21(6):1087–1092

Mill JS (2011) A system of logic, ratiocinative and inductive (Classic Reprint). Oxford University Press, Oxford

Mistry J, Nelwamondo FV, Marwala T (2006) Investigating demographic influences for HIV classification using Bayesian autoassociative neural networks. Lecture notes in computer science vol. 5507, pp 752–759

Moore GE (1965) Cramming more components onto integrated circuits. Electronics 38(8):114–117

Mthembu L, Marwala T, Friswell MI, Adhikari S (2010) Finite element model selection using particle swarm optimization Conference Proceedings of the Society for Experimental Mechanics Series, 1, vol 13, Dynamics of Civil Structures, vol 4, Springer London, pp 41-52 Tom Proulx (ed), 2010

Murata A, Kubo S, Hata N (2012) Study on promotion of cooperative behavior in social dilemma situation by introduction of bounded rationality-Effects of group heuristics on cooperative behavior. Proceedings of the SICE annual conference, (art. no. 6318444), pp 261–266

Muth JF (1961) Rational expectations and the theory of price movements. Econometrica 29(3):315-335. doi:10.2307/1909635. Reprinted In: Hoover KD, (ed) (1992) The new classical macroeconomics, vol. 1

Neal RM (1992) Bayesian training of back-propagation networks by the hybrid Monte Carlo Method. Technical Report CRG-TR −92-1, Department of Computer Science, University of Toronto

Nelwamondo FV, Marwala T (2007) Handling missing data from heteroskedastic and nonstationary data. Lecture notes in computer science vol 4491(1). Springer, Berlin, pp 1297–1306

Nozick R (1993) The nature of rationality. Princeton University Press, Princeton

O'Rourke P, Josephson J (eds) (1997) Automated abduction: inference to the best explanation. AAAI Press, California

Palma G, Bia P, Mescia L, Yano T, Nazabal V, Taguchi J, Moréac A, Prudenzano F (2014) Design of fiber coupled Er3+ :chalcogenide microsphere amplifier via particle swarm optimization algorithm. Opt Eng 53(7), (art. no. 071805), pp 1–8

Pantanowitz A, Marwala T (2009) Evaluating the impact of missing data imputation. Advanced data mining and application 5th international conference, ADMA 2009, Lecture notes in computer science vol. 5678. Springer Berlin Heidelberg, pp. 577–586

Patel P, Marwala T (2006) Neural networks, fuzzy inference systems and adaptive-neuro fuzzy inference systems for financial decision making. 13th international conference, ICONIP, Neural information processing, Lecture notes in computer science vol. 4234. Springer Berlin Heidelberg, pp 430–439

Psillos S (2009) An explorer upon untrodden ground: peirce on abduction. In: Gabbay DM, Hartmann S, Woods J (eds) Handbook of the history of logic 10. Elsevier. pp 117–152

Russell SJ, Norvig P (2003) Artificial intelligence: a modern approach, 2nd edn. Prentice Hall, Upper Saddle River

Sadeghi J, Sadeghi S, Niaki STA (2014) Optimizing a hybrid vendor-managed inventory and transportation problem with fuzzy demand: an improved particle swarm optimization algorithm. Inf Sci 272:126–144

Sargent TJ (1987) Rational expectations. The new Palgrave dictionary of economics, 2nd edn, vol. 4. Palgrave Macmillan, Basingstoke, pp 76–79

Savin NE (1987) Rational expectations: econometric implications. The new Palgrave dictionary of economics, 2nd edn, vol. 4. Palgrave Macmillan, Basingstoke, pp 79–85

Sen A (2008) Rational behavior. The new Palgrave dictionary of economics, 2nd edn, vol. 4. Palgrave Macmillan, Basingstoke, pp 76–79

Simon H (1957) A behavioral model of rational choice. In: Models of man, social and rational: mathematical essays on rational human behavior in a social setting. Wiley, New York

Simon H (1991) Bounded rationality and organizational learning. Organ Sci 2(1):125–134

Snowdon B, Vane H, Wynarczyk P (1994) A modern guide to macroeconomics. An introduction to competing schools of thought. Cambridge, Edward Elgar Publishing Limited, pp 236–279, ISBN: 1-85278-884-4 HB

Spohn W (2002) The many facets of the theory of rationality. Croat J Philos 2(3):247–262

Stanciu-Viziteu LD (2012) The shark game: equilibrium with bounded rationality. Managing market complexity. The approach of artificial economics. Lecture notes in economics mathematical systems, vol. 662. Springer Berlin Heidelberg, pp 103–111

Takaishi T (2014) Bayesian estimation of realized stochastic volatility model by hybrid Monte Carlo algorithm. J Phys Conf Ser 490(1), (art. no. 012092), arXiv:1408.0981

Tisdell C (1996) Bounded rationality and economic evolution: a contribution to decision making, economics, and management. Edward Elgar Publishing Limited, Cheltenham

Tsang EPK (2008) Computational intelligence determines effective rationality. Intl J Autom Control 5(1):63–66

Valdez F, Melin P, Castillo O (2014) Modular neural networks architecture optimization with a new nature inspired method using a fuzzy combination of particle swarm optimization and genetic algorithms. Inf Sci 270:143–153

Vyse SA (2000) Believing in magic: the psychology of superstition. Oxford University Press, Oxford

Widom M, Huhn WP, Maiti S, Steurer W (2014) Hybrid Monte Carlo/molecular dynamics simulation of a refractory metal high entropy alloy. Metall Mater Trans 45(1):196–200

Yao J, Li D (2013) Bounded rationality as a source of loss aversion and optimism: a study of psychological adaptation under incomplete information. J Econ Dyn Control 37(1):18–31

Chapter 7
Filtering Irrelevant Information for Rational Decision Making

7.1 Introduction

One aspect of rational decision making is that it ought to use relevant information. For example, if one wanted to decide whether to go to McDonalds and order a burger, then he first jumps into the pool for no reason other than the fact that he thinks it is a necessary step to decide to go to McDonalds and order a burger we would characterize him as being irrational. The problem of abundance of information makes the task of decision making difficult because it then requires a great deal of competence to separate irrelevant information from relevant information. In his seminal book, Marwala (2014) describes the theory of marginalization of irrationality in decision making in order to make satisfying decisions in the presence of irrationality. This he achieved by assuming that decision making process is made of a series of decision making actions which are either rational or irrational. The aggregation of these actions determine whether an irrational decision process satisfies or not. He applied this theory for the diagnosis of breast cancer and observed that the more irrational a decision making process the less accurate the results become. He also observed that indeed it is possible for irrational actions to be marginalized by rational actions and that irrationality can also marginalize rationality thus resulting in irrational decision making. Another techniques for dealing with irrelevant information in rational decision making include the work by Starfield et al. (2006, 2007) who demonstrated that near-field artifacts and ghosting can be reduced geometrically by placing multiple identical limited-field-of-view coded apertures side by side, in the form of a matrix. The results they obtained showed that coded apertures introduced a new 'tartan' artifact, but that the method of limiting the field-of-view controls the 'tartan' and thus reduce near-field artifacts and ghosting are also reduced. This then allows radiologists to make better decisions by minimizing irrelevant information.

On decision making irrationality always exists whether in the nature of the data that is used to make a decision or in the manner in which the data is processed. For example, Shen (2014) studied the interplay between consumer rationality/ irrationality and its implication on financial literacy in the credit card market.

© Springer International Publishing Switzerland 2014
T. Marwala, *Artificial Intelligence Techniques for Rational Decision Making*,
Advanced Information and Knowledge Processing,
DOI 10.1007/978-3-319-11424-8_7

Another study by Dayan (2014) showed how irrational choice can be rationalized by using systematic statistical approaches. Hess and Orbe (2013) studied irrationality in survey forecast and its implications on the anchoring bias test whereas Chohra et al. (2013) studied the impact of irrationality on negotiation strategies with incomplete data.

This chapter studies how to handle irrelevant information in order to make rational decision. As already explained, it is irrational to use irrelevant information to make a decision. The first technique which is studied is the theory of marginalization of irrationality for decision making. In this model information that are relevant and irrelevant are used together and relevant information crowds out irrelevant information. In this chapter, this technique is applied using radial basis functions for credit scoring.

The second theory that is studied is the automatic relevance determination method where the Bayesian method is used to score the relevance of each variable for decision making (Shutin and Buchgraber 2012; Tan and Févotte 2013). Ayhan et al. (2013) successfully applied composite kernels to automatically determine the anatomical regions of relevance for diagnosis of Alzheimer's disease whereas Duma et al. (2012) successfully applied the automatic relevance determination for modeling missing data. Yamaguchi (2012) successfully applied variational Bayesian inference for generative topographic mapping.

The third technique which is used to deal with irrelevant information is the principal component analysis (PCA). This procedure uses the eigenvalue analysis to transform the data into its principal components which are orthogonal (Idrich 2006; Strang 1993). Then the variances of the transformed data along the principal components are analyzed and the components with insignificant variance are ignored and by so doing the irrelevant information is ignored. Yan et al. (2014) successfully applied PCA for aberration measurement of aerial images whereas Karaman et al. (2014) successfully applied PCA to characterize rheological properties of hydrocolloids. Wang et al. (2014) successfully applied principal component and causal analysis in vitro toxicity data for nanoparticles. Preisendorfer (1988) studied the principal component analysis and applied it to meteorology and oceanography whereas Pegram (1983) used PCA to study spatial relationships in stream flow residuals by describing precipitation and climatically dependent hydrologically homogeneous regions on a continent.

The fourth method used is the independent component analysis (ICA) which is a method of breaking a signal into independent components. By so doing one is able to focus on one component of a signal which is hopefully relevant for decision making. Yang and Nagarajaiah (2014) successfully applied ICA for blind identification of damage in time-varying systems whereas Chen et al. (2014b) applied successfully ICA in financial risk management and was found to perform better than the PCA. Metsomaa et al. (2014) applied the ICA successfully for multi-trial evoked EEG thereby identifying hidden components.

These techniques described above are applied in this chapter to problems of condition monitoring, credit scoring, interstate conflict, blind source separation

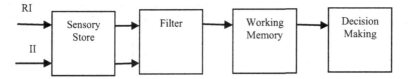

Fig. 7.1 Bounded rational decision making framework based on Broadbent model. Key: *RI* Relevant Information, *II* Irrelevant Information

and detection of epileptic activity. The next section describes the cocktail problem which is a problem that studies how a person in a cocktail party is able to filter other voices and concentrates on one voice.

7.2 Cocktail Party Problem: How People Deal With Irrelevant Information?

This section describes a problem which is called the cocktail problem (Hafter et al. 2013; Schmidt and Römer 2011; Collins 2011; Cohen 2006; Bronkhorst 2000; Wood and Cowan 1995; Cherry 1953). This problem is about a person who is in a loud cocktail party but he is able to only listen to one person. How does he do that? How does he filter all the noise in the room and only focus on one person? If the conversation is recorded and he listened to it later he will have difficulty filtering. Psychological experiments have shown that when a person has only one ear, it is difficult to separate the voice to focus on amidst all the background noise (Hawley et al. 2004).

There are a large number of models that have been proposed to explain the cocktail problem and these include the Broadbent model, Treisman model and the Kahneman model (Broadbent 1954; Treisman 1969; Kahneman 1973; Scharf 1990; Brungart and Simpson 2007; Haykin and Chen 2005). This chapter describes the Broadbent and Treisman models within the context of bounded rational decision making.

The Broadbent model for bounded rational decision making is described in Fig. 7.1. The Broadbent model is also called the filter model and is premised on the fact that people are more capable of recalling information that they actively attended to than the information that they did not attend to, and therefore there must be a filter that blocks information that they did not attend to. In this figure two types of information are the input to bounded rational decision making device and these are relevant information (RI) and irrelevant information (II). The information enters the body through sensory organs such as ears, eyes etc. and then are stored in the sensory memory. Then the information is passed through the selective filter which only allows information that was attended to and in the present context only relevant information passes through the filter and is stored in the working memory to be used for decision making.

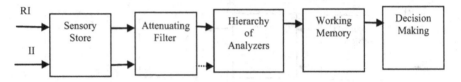

Fig. 7.2 Decision making framework based on Tresisman model

The other model which is an extension of the Broadbent model is the Treisman model which is described in Fig. 7.2. The Treisman model differs from the Broadbent model in that the filter does not block the irrelevant information but assign weighting factors and the hierarchy of analyzers and select the information to be put in the working memory for decision making.

7.3   Marginalization of Irrationality Theory

This chapter applies the theory of the marginalization of irrationality for effective decision making as was done in Marwala (2014) in the knowledge that in decision making rationality offers advantages over irrationality. This theory of the marginalization of irrationality for decision making states that when making a complex decision which contains rational and irrational decision actions, that decision will be effective or workable if the irrational actions are subdued by the rational actions. To demonstrate the marginalization of irrationality theory, we use the principle of signal to noise ratio that has been widely applied in information theory (Schroeder 1999; Choma et al. 2003; Russ 2007; Raol 2009). The principle of signal to noise ratio is used to assess the degree through which the signal is corrupted or influenced by noise. When the noise overwhelms or marginalizes the signal, then the signal cannot transmit the relevant information effectively while when the signal marginalizes the noise then the signal is able to transmit a significant part of the original information. By the same token when the rational aspect of a decision making process marginalizes the irrational aspect, then the consequence of the decision is effective or workable while when the irrational aspect of decision making marginalizes the rational aspect, then the decision is ineffective or unworkable. However, this problem can be dealt with using principal component analysis and independent component analysis which is described later in the chapter.

The rational (wanted actions) to irrational (unwanted actions) ratio (RIR) is defined as follows (Schroeder 1999; Choma et al. 2003; Russ 2007; Marwala 2014):

$$RIR = \frac{P_{rational}}{P_{irrational}} \tag{7.1}$$

Here $P_{rationality}$ is the impact of the rational actions whereas $P_{irrationality}$ is the impact of the irrational actions. When RIR is high then the rational actions marginalize irrational actions and therefore the decision is *satisffective* meaning that the outcome of the decision making process is satisfactory and effective while a low *RIR* simply means an unsatisfactory and/or ineffective decision making. The steps that are followed to make a decision when the process is not rational are as follows (Marwala 2014):

1. Collect the information needed to make a decision.
2. Outline a process to be followed to make such as a decision.
3. Divide the decision making process into sets of actions with each action classified into whether it is rational or irrational.
4. Assess whether irrational actions of the process are marginalizable when compared to the rational actions.
5. If the irrational actions are marginalizable then the solution is satisffective whereas if these actions are not marginalizable then the solution is not satisffective.

Clearly satisffecting an irrational decision requires the assessment as to whether the irrational aspect of an irrational process of decision making is marginalizable. The theory of marginalization of information is also applicable to any decision making process where the information that is needed to make such a decision is not complete and there is no instrument of completing fully the missing information. In this situation, the decision maker simply ignores or marginalizes the missing information and continues to make a decision. The efficacy of this decision will depend on the ratio of the power of observed information to the power of the missing information just exactly as is the case in information theory where the usefulness of the information depends on the signal to noise ratio.

We apply the theory of the marginalization of irrationality for credit scoring. To achieve this we use the radial basis function (RBF) network (Rad et al. 2014; Link et al. 2014). The RBF network is mathematically written as follows (Bishop 1995; Malek-Khatabi et al. 2014):

$$y_k(\{x\}) = \sum_{j=1}^{M} w_{jk} \exp\left(-\beta(\{x\}) - \{c\}_j\right)^2 \tag{7.2}$$

Here w_{jk} denotes the output weights, each corresponding to the connection between a hidden unit and an output unit, M denotes the number of hidden units, $\{c\}_j$ is the center for the jth neuron, $\phi_j(\{x\})$ is the jth non-linear activation function, $\{x\}$ the input vector, and $k = 1, 2, 3, \ldots, M$ (Bishop 1995). The choice of the number of hidden nodes, M, is part of the model selection process. Optimization approaches such as genetic algorithms, simulated annealing, and particle swarm optimization have been applied to choose the optimal RBF design (Marwala 2009). The activation selected in this chapter for the hidden layers is the Gaussian distribution function (Marwala 2009; Bishop 1995). The RBF has weights in the outer layer only, while the hidden nodes have what are called the centers. Training an RBF network involves identifying two sets of parameters: the centers and the output weights which can both be looked at as free parameters in a regression framework.

The centers and network weights can both be identified concurrently using the full Bayesian method or can then again be identified using the maximization of expectation procedure (Bishop 1995; Marwala 2009). In this chapter, a two-phase training process was used to first identify the centers followed by a determination of the network weights. The first phase is to use a self-organizing map, the so-called the *k-means* clustering technique, to determine the centers. The step of identifying the centers only contemplates the input space, while the identification of the network weights studies both the input and output space.

The *k-means* technique is intended to group objects based on attributes into k subdivisions. For the RBF, k is equal to the number of centers, M. The objective of the *k-means* algorithm is to identify the centers of natural clusters in the data and to assume that the object attributes form a vector space. This is achieved by minimizing the total intra-cluster variance, or, the squared error function which can be expressed as follows (Hartigan 1975; Hartigan and Wong 1979; Marwala 2009):

$$E = \sum_{i=1}^{C} \sum_{x_j \in S_i} \left(\{x\}_j - \{c\}_i \right)^2 \tag{7.3}$$

Here, C is the number of clusters in S_i, $i = 1, 2, \ldots, M$ and $\{c\}_i$ is the center of the ith point $x_j \in S_i$. Lloyd algorithm (Lloyd 1982) is applied to find the cluster centers. Lloyd algorithm is initialized by randomly dividing the input space into k initial sets or using heuristic data. The mean point is calculated for each set and then a new partition is built by associating each point with the closest center. The centroids are then re-calculated for the new clusters, and the process is repeated by changing these two steps until convergence. Convergence is attained when the centroids no longer change or the points no longer switch clusters.

The network weights are identified given the training data and the estimated centers using the Moore-Penrose pseudo inverse (Moore 1920; Penrose 1955; Ben-Israel and Greville 2003). The approximation of the network weights given the centers is a linear process. The RBF equation can be written as follows after the centers have been identified:

$$[y_{ij}] = [\phi_{ik}][w_{kj}] \tag{7.4}$$

Here, $[y_{ij}]$ is the output matrix, with i representing the number of training examples and j representing the number of outputs. Parameter $[\phi_{ik}]$ is the activation function matrix in the hidden layer, i represents the training examples, k is the number of hidden neurons while $[w_{kj}]$ is the weight matrix. The weight matrix $[w_{kj}]$ can be identified by inverting the activation function matrix $[\phi_{ik}]$ if it is a square matrix. Nevertheless, this matrix is not square and, consequently, it cannot be inverted using standard matrix methods. Fortunately, this matrix can be inverted using the Moore-Penrose pseudo-inverse technique which is expressed mathematically as follows (Penrose 1955):

$$[\phi_{ik}]^* = \left([\phi_{ik}][\phi_{ik}]^T \right)^{-1} [\phi_{ik}]^T \tag{7.5}$$

This, consequently, indicates that the weight matrix may be approximated by using the pseudo inverse matrix as follows (Marwala 2009):

$$[w_{kj}] = [\phi_{ik}]^*[y_{ij}] \tag{7.6}$$

The concept of marginalization of irrelevant information theory is tested on the Australian credit coring data. More information on this is found in Bache and Lichman (2013). This data set has been used previously by many researchers (Setiono and Liu 1997). This dataset has 14 attributes which are used to predict a credit score of a customer. The 14 attributes were 6 numeric and 8 categorical. The RBF network was constructed with 14 input units with one output unit and 8 hidden units in the one hidden layer. The network was trained with 230 examples, the model was selected using 230 examples and then tested using 230 examples and the results are shown in Fig. 7.3. When all the 14 input were used, the RBF achieved an accuracy of 91 %. When one of the variables were randomly generated thus making the variable irrelevant, an accuracy of 90 % was achieved.

These results show that the irrelevant variable was marginalized by the other 13 variables. This process is repeated and as shown in Fig. 7.3, the relevant data marginalizes the irrelevant variables. This of course can just mean that the data variables are highly correlated

7.4   Automatic Relevance Determination Theory

The next technique that is used to marginalize irrelevant information is the automatic relevance determination (ARD) proposed by MacKay (1991). The ARD basically ranks the input data according to the degree of importance as far as its ability to predict the output is concerned. The ARD framework is shown in Fig. 7.4.

Naturally the ARD will give a low ranking for irrelevant inputs. In this section we apply the multi-layer perceptron (MLP) to implement the ARD method (Xu et al. 2014; Abbasi and Eslamloueyan 2014; Ahmad 2014; Pazoki et al. 2014; Islam et al. 2014). The relationship between output, y, and input variables, x, for the MLP may be written as follows (Bishop 1995):

$$y_k = f_{outer}\left(\sum_{j=1}^{M} w_{kj}^{(2)} f_{inner}\left(\sum_{i=1}^{d} w_{ji}^{(1)} x_i + w_{j0}^{(1)}\right) + w_{k0}^{(2)}\right) \tag{7.7}$$

Here, $w_{ji}^{(1)}$ and $w_{ji}^{(2)}$ indicate neural network weights in the first and second layers, respectively, going from input i to hidden unit j, M is the number of hidden units, d is the number of output units while $w_{j0}^{(1)}$ indicates the bias for the hidden unit j. If the training set $D = \{x_k, y_k\}_{k=1}^{N}$ is used, where superscript N is the number of training examples, and assuming that the targets y are sampled independently given the kth inputs x_k and the weight parameters w_{kj} then the cost function, E, may be written using the cross-entropy cost function as follows (Bishop 1995):

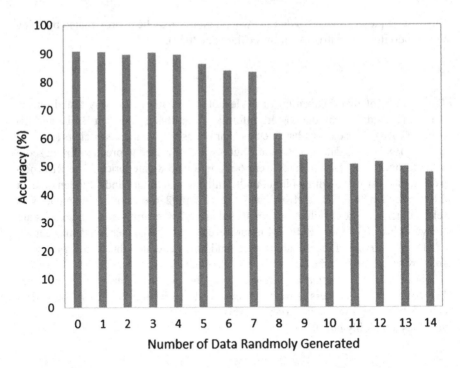

Fig. 7.3 Prediction accuracy for a given number of randomly generated data

Fig. 7.4 Decision making framework based on ARD model

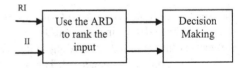

$$E_D = -\beta \sum_{n=1}^{N} \sum_{k=1}^{K} \{t_{nk} \ln(y_{nk}) + (1 - t_{nk}) \ln(1 - y_{nk})\} \qquad (7.8)$$

Here, t_{nk} is the target vector for the nth output and k^{th} training example, N is the number of training examples, K is the number of network output units, n is the index for the training pattern, b is the data contribution to the error, and k is the index for the output units.

Automatic relevance determination is a procedure where in a predictive model the causal variable is ranked in order of its effect on the output variable. In this case if there is no causal relationship between variables a and b then variable a will be assigned an irrelevancy factor. The ARD technique is created by associating the hyper-parameters of the prior with each input variable and, therefore, the prior can be generalized as follows (MacKay 1991, 1992):

$$E_W = \frac{1}{2} \sum_{k} \alpha_k \{w\}^T [I_k] \{w\} \qquad (7.9)$$

where superscript T is the transpose, k is the weight group and $[I]$ is the identity matrix. Given the generalized prior in eq. 7.9, the posterior probability then is (Bishop 1995):

$$P(\{w\}\,|\,[D], H_i)$$

$$= \frac{1}{Z_s} \exp\left(\beta \sum_n \{t_n - y(\{x\}_n\}^2 - \frac{1}{2} \sum_k \alpha_k \{w\}^T [I_k]\{w\} \right) \tag{7.10}$$

here

$$Z_E(\alpha, \beta) = \left(\frac{2\pi}{\beta}\right)^{N/2} + \prod_k \left(\frac{2\pi}{\alpha_k}\right)^{W_k/2} \tag{7.11}$$

here W_k is the number of weights in group k. The evidence can be written as follows (Bishop 1995):

$$p([D]|\alpha, \beta) = \frac{1}{Z_D Z_W} \int \exp(-E(\{w\})) d\{w\}$$

$$= \frac{Z_E}{Z_D Z_W}$$

$$= \frac{\left(\dfrac{2\pi}{\beta}\right)^{N/2} + \prod_k \left(\dfrac{2\pi}{\alpha_k}\right)^{W_k/2}}{\left(\dfrac{2\pi}{\beta}\right)^{N/2} \prod_k \left(\dfrac{2\pi}{\alpha_k}\right)^{W_k/2}} \tag{7.12}$$

Maximizing the log evidence gives the following estimations for the hyper-parameters (Bishop 1995):

$$\beta^{MP} = \frac{N - \gamma}{2E_D(\{w\}^{MP})} \tag{7.13}$$

$$\alpha_k^{MP} = \frac{\gamma_k}{2E_{W_k}(\{w\}^{MP})} \tag{7.14}$$

here $\gamma = \sum_k \gamma_k$, $2E_{W_k} = \{w\}^T I_k \{w\}$ and

$$\gamma_k = \sum_j \left(\frac{\pi_j - \alpha_k}{\eta_j} ([V]^T [I_k][V])_{jj} \right) \tag{7.15}$$

and $\{w\}^{MP}$ is the weight vector at the maximum point and this is identified in this chapter using the scaled conjugate gradient method, η_j are the eigenvalues of $[A]$, and $[V]$ are the eigenvalues such that $[V]^T[V] = [I]$.

The relevance of each input variable, the α_k^{MP} and β^{MP} is estimated by choosing the initial values for the hyper-parameters randomly, training the network using scaled conjugate gradient method to identify $\{w\}^{MP}$ and using eq. 7.13 and 7.14 to estimate the hyper parameters and repeating the procedure until convergence without further initialization of the hyper parameters (MacKay 1991). When the α is low then the relevance of a variable on predicting the output variable is high. The MLP trained with the ARD framework automatically marginalizes irrelevant variables depending on how relevant these variables are.

The ARD approach to training neural networks is tested to model interstate problem and the description below is from the work by Marwala and Lagazio (2011). On implementing ARD for conflict prediction, four variables associated with realist analysis and three "Kantian" variables are used. The first variable is *Allies*, a binary measure coded 1 if the members of a dyad are linked by any form of military alliance, and 0 in the absence of military alliance. *Contingency* is also binary, coded 1 if both states share a common boundary and 0 if they do not, and *Distance* is the logarithm, to the base 10, of the distance in kilometers between the two states' capitals. *Major Power* is a binary variable, coded 1 if either or both states in the dyad is a major power and 0 if neither are super powers. *Capability* is the logarithm, to the base 10, of the ratio of the total population plus the number of people in urban areas plus industrial energy consumption plus iron and steel production plus the number of military personnel in active duty plus military expenditure in dollars in the last 5 years measured on stronger country to weak country. The variable *Democracy* is measured on a scale where the value of 10 is for an extreme democracy and a value of -10 is an extreme autocracy and taking the lowest value of the two countries. The variable *Dependency* is measured as the sum of the countries import and export with its partner divided by the Gross Domestic Product of the stronger country. It is a continuous variable measuring the level of economic interdependence (dyadic trade as a portion of a state's gross domestic product) of the less economically dependent state in the dyad. These measures were derived from conceptualizations and measurements conducted by the Correlates of War (COW) project.

The COW data are used to generate training and testing sets. The training data set consists of 500 conflicts and 500 non-conflict cases, and the test data consists of 392 conflict data and 392 peace data. We use a balanced training set, with a randomly selected equal number of conflicts and non-conflicts cases, to produce robust classifications and stronger insights on the reasons of conflicts. The data are normalized to fall between 0 and 1. This is done to improve the effectiveness of neural networks modeling (Bishop 1995). The MLP architecture was chosen has $M=10$, a logistic function in the output layer, and a hyperbolic function in the hidden layers as the optimal architecture. When the ARD is implemented to train the network and the hyperparameters calculated and then the inverse of the hyperparameters is calculated, and the results are shown in Fig. 7.5 and this results are reported in Marwala and Lagazio (2011). Figure 7.5 indicates that the *Dependency* variable has the highest relevance, followed by *Capability*, followed by *Democracy* and then *Allies*. The

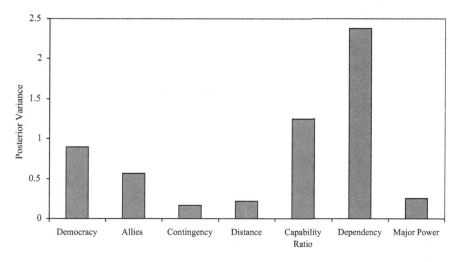

Fig. 7.5 Relevance of each liberal variable with regards to the classification of MIDs. (Marwala and Lagazio 2011)

Table 7.1 Classification results

Method	True conflicts TC	False peaces FP	True peaces TP	False conflicts FC
ARD MLP	298	94	302	90
MLP	295	97	299	93

remaining three variables, that is, *Contingency*, *Distance*, and *Major Power*, have similar impact although it is much smaller in comparison with the other two liberal variables, democracy and economic interdependence, and the two realist variables, allies and difference in capabilities.

The results obtained when the ARD based MLP is used to predict conflict are shown in Table 7.1. These results indicate that the MLP based on ARD performs better than the regular MLP. This is because relevance is factored into account in the ARD based MLP as compared to a regular MLP.

7.5   Principal Component Analysis

One other technique for factoring out irrelevant information is the principal component analysis (PCA). The PCA is implemented to reduce the input data into independent components (Alfaro et al. 2014; Jolliffe 2002; Marwala 2001; Karhunen 1947). The PCA uses the eigenvalue analysis to orthogonalizes the components of the input vector so that they are uncorrelated with each other. When implementing the PCA for data reduction, correlations and interactions among variables in the data are summarised in terms of a small number of underlying factors that are in the form of eigenvalues and eigenvectors. Pearson (1901) introduced the PCA to

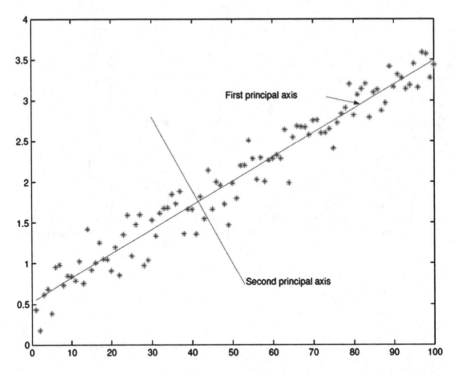

Fig. 7.6 This figure illustrates how a 2 dimensional figure is transformed into one dimension by calculating the eigenvectors and transforming the data into two principal axes and ignoring the second principal component. (Marwala 2001)

recast linear regression analysis into a new framework, and was advanced further by Hotelling (1933) and generalized by Loéve (1978). The PCA has been successfully applied to reduce the dimension of the data (Bishop 1995) and has been success-fully used in a variety of applications such as in characterization of rheological and physicochemical properties (Karaman et al. 2014), in tracking the curing process of automotive paint (Chen et al. 2014a), and characterisation of organic and conventional sweet basil (Lu et al. 2014).

The PCA identifies the directions in which the data points have the most vari-ance and these directions are called *principal directions*. The data are then projected onto these principal directions without the loss of significant information from the data. An illustration of the PCA is shown in Fig. 7.6. Here follows a brief outline of the implementation of the PCA. The covariance matrix is calculated as follows (Marwala 2001):

$$[\Sigma] = \sum_{i=1}^{N} (\{x\}_i - \mu)(\{x\}_i - \mu)^T \qquad (7.16)$$

Here Σ is the covariance matrix, N is the number of training examples, $\{x\}_i$ is the vector which corresponds to a training example and when these training examples

Table 7.2 Fault cases used to train, cross-validate and test the networks. (Marwala 2001)

Fault	[000]	[100]	[010]	[001]	[110]	[101]	[011]	[111]
Training set	21	21	21	21	21	21	21	21
Test set	39	3	3	3	3	3	3	39

are many they form a matrix $[x]$, and μ is the mean vector of the data set taken over the number of training set. The eigenvalues $[\lambda]$ and eigenvectors $[\Phi]$ of the covariance matrix can be written as follows:

$$[\Sigma][\Phi] = [\lambda][\Phi] \tag{7.17}$$

There are many methods that have been proposed to estimate these eigenvalues and eigenvectors such as the Gerschgorin method (Gerschgorin 1931; Varga 2002; Golub and Van Loan 1996). After the eigenvalues and the eigenvectors have been estimated then the eigenvalues are compared and the eigenvalues which are numerically high correspond to eigenvectors with high variance. To reduce the dimension of the data $[x]$ we ignore eigenvectors corresponding to eigenvalues which have low numerical values thus forming a reduced eigenvector matrix $[\Phi_R]$. Thus $[x]$ can be reduced into lower dimension as follows:

$$[x_R] = [\Phi_R][x] \tag{7.18}$$

The PCA is applied in the problem to condition monitoring of structures where vibration data are used to predict whether the structure is healthy or faulty. This subject was treated appropriately by Marwala (2012). Multi-fold cross-validation technique was used to train and validate the MLP neural network based on modal properties data for decision making (Marwala 2001). The fault cases that were used to train and test the networks are shown in Table 7.1. If all the modal properties identified were used to train the neural network, then this neural network will have 340 inputs. The idea of using 340 variables to make a decision is a complicated and daunting task. In line with the principle of Occam's Razor which states that the correct model is the simplest model that describes the evidence, these 340 variables from the modal properties ought to be reduced and in this chapter as was done by Marwala (2001) they were reduced to 10 using the PCA. The reduction of this data to this low dimension is mainly done by eliminating information that is not essential in decision making. Making decisions using information which is irrelevant is not rational. So the application of the PCA which essentially eliminates irrelevant information is a rational action. (Table 7.2)

The MLP neural network had 10 inputs, 9 hidden nodes and one output. More details on this network are found in Marwala (2001). The confusion matrix obtained when the MLP and modal properties were used is shown in Table 7.3. This table demonstrates that this network classifies 89.7% of [000] fault cases correctly; all one and two-fault cases with the exception of three [101] cases correctly; and 71.8% of [111] fault cases correctly. Of the four [000] cases that were classified wrongly by

Table 7.3 Confusion matrix from the classification of fault cases in the test data using the MLP network and modal properties

		Predicted							
		[000]	[100]	[010]	[001]	[110]	[101]	[011]	[111]
Actual	[000]	35	0	1	3	0	0	0	0
	[100]	0	3	0	0	0	0	0	0
	[010]	0	0	3	0	0	0	0	0
	[001]	0	0	0	3	0	0	0	0
	[110]	0	0	0	0	3	0	0	0
	[101]	0	0	0	3	0	0	0	0
	[011]	0	0	0	0	0	0	3	0
	[111]	0	0	0	0	8	1	2	28

the network, one is classified as a [010] case and three as [001] cases. Of the eleven [111] cases that were classified wrongly by the network, eight were classified as [110] cases, one as a [101] case and two as [011] cases. The three [101] cases that were misclassified by the network were all classified wrongly as [001] cases.

7.6 Independent Component Analysis: Blind Source Separation

This section describes a methodology called the independent component analysis which is a method of separating multi-variable data into components which are independent component analysis (ICA) (Stone 2004; Hyvarinen et al. 2001). Zhan et al. (2014) successfully applied ICA in dam deformation analysis whereas Palmieri et al. (2014) successfully applied the ICA for network anomaly detection.

Ding et al. (2014) successfully applied ICA for image quality assessment whereas Zhang et al. (2014) successfully applied the ICA for source contribution estimation and Fan and Wang (2014) applied successfully the ICA for fault detection and diagnosis of non-linear non-Gaussian dynamic processes. Suppose two speeches as x_1 and x_2 and measurements are made at z_1 and z_2 as shown in Fig. 7.7 then a relationship can be written which relates x_1 and x_2 as well as z_1 and z_2 as follows:

$$[A]\begin{Bmatrix} x_1 \\ x_2 \end{Bmatrix} = \begin{Bmatrix} z_1 \\ z_2 \end{Bmatrix} \tag{7.19}$$

Here $[A]$ is a vector and for this case it is a 2 by 2 matrix. This equation can be rewritten as follows where $[W] \approx [A]^{-1}$:

$$\begin{Bmatrix} x_1 \\ x_2 \end{Bmatrix} = [W]\begin{Bmatrix} z_1 \\ z_2 \end{Bmatrix} \tag{7.20}$$

Fig. 7.7 Two speech signals being integrated and then separated using ICA

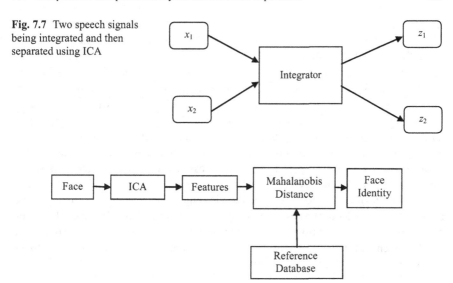

Fig. 7.8 Face recognition algorithm based on the ICA and Mahalanobis distance

Given the fact that x_1 and x_2 are independent, $[W]$ matrix can be estimated by ensuring that the resulting x_1 and x_2 are as independent as possible. By applying the maximum likelihood approach the Likelihood of $[W]$ can be written as follows (Stone 2004):

$$L([W]) = p_s([W]\{z\})[W] \tag{7.21}$$

Here p_s is the probability distribution function and if we assume the high Kurtosis model then the objective function which maximizes independence of variables z_1 and z_2 can be expressed as follows.

$$\ln L([W]) = \frac{1}{N}\sum_{i}^{M}\sum_{j}^{N}\ln(1 - \tanh(w_i^T z_j)) + \ln([W]) \tag{7.22}$$

The ICA has been used for a human face recognition by Surajpal (2007) as well as Surajpal and Marwala (2007). The procedure employed in this chapter is based on the work of Surajpal and is shown in Fig. 7.8.

This figure indicates that the data from the pixels of the face is separated into independent components. These features are compared to the features in the database using the Mahalanobis distance to identify the face. The Mahalanobis distance between two vector $\{x\}$ and $\{y\}$ can be written as follows (Mahalanobis 1936):

$$d(\{x\},\{y\}) = \sqrt{(\{x\}-\{y\})^T S^{-1}(\{x\}-\{y\})} \tag{7.23}$$

Here S is the covariance matrix.

As described by Gross et al. (2001) as well as Surajpal (2007) the most influential factors that unfavorably affect recognition rates are viewing angle, illumination, expression, occlusion and time delay. As was done by Surajpal (2007) the face database (Martinez and Benavente 1998) which contained over 3000 images of 135 individuals (76 males and 60 females) were used. The data set was gathered over two sessions with a two week interval between shoots and 13 images per subject under varying conditions with one neutral and 3 under different expression, illumination and occlusion (with lighting changes). Each image was cropped and resized to dimensions and pre-processed. The preprocessed database formulated a baseline system which was then divided into gallery and probe sets. 110 subjects (55 males and 55 females) were randomly selected, yielding a database of 2860 images. The method was tested and the results indicated a recognition of 72%. If the ICA was not used then the recognition rate of 65% was achieved. The ICA features only deal with independent data and therefore avoids redundant information which confuses the system due to a higher complexity.

Conclusions

This chapter studied methods that are used for dealing with irrelevant information. Four techniques were considered and these were the marginalization of irrationality approach, automatic relevance determination, principal component analysis and independent component analysis. These techniques were applied to the problems of condition monitoring, credit scoring, interstate conflict and face recognition and were found to be effective on handling irrelevant information.

References

Abbasi A, Eslamloueyan R (2014) Determination of binary diffusion coefficients of hydrocarbon mixtures using MLP and ANFIS networks based on QSPR method. Chemom Intell Lab Syst 132:39–51

Ahmad I (2014) Enhancing MLP performance in intrusion detection using optimal feature subset selection based on genetic principal components. Applied Mathematics and. Inf Sci 8(2):639–649

Alfaro CA, Aydin B, Valencia CE, Bullitt E, Ladha A (2014) Dimension reduction in principal component analysis for trees. Comput Stat Data Anal 74:157–179

Ayhan MS, Benton RG, Raghavan VV, Choubey S (2013) Composite kernels for automatic relevance determination in computerized diagnosis of Alzheimer's disease. Lecture Notes in Computer Science (including subseries Lecture Notes in Artificial Intelligence and Lecture Notes in Bioinformatics). LNAI 8211:126–137

Bache K, Lichman M (2013) UCI Machine Learning Repository. [http://archive.ics.uci.edu/ml]. Irvine, CA: University of California, School of Information and Computer Science

Ben-Israel A, Greville TNE (2003) Generalized inverses. Springer, New York

Bishop CM (1995) Neural networks for pattern recognition. Oxford University Press, Oxford

Broadbent DE (1954) The role of auditory localization in attention and memory span. J Exp Psychol 47(3):191–196

Bronkhorst AW (2000) The cocktail party phenomenon: a review on speech intelligibility in multiple-talker conditions (pdf). Acta Acust United Acust 86:117–128

Brungart DS, Simpson BD (2007) Cocktail party listening in a dynamic multitalker environment. Percept Psychophys 69(1):79–91

Chen J-B, Sun S-Q, Yu J, Zhou Q (2014a) Tracking the curing process of automotive paint by moving-window two-dimensional infrared correlation spectroscopy and principal component analysis. J Mol Struct 1069(1):112–117

Chen R-B, Chen Y, Härdle WK, (2014b) TVICA-Time varying independent component analysis and its application to financial data. Comput Stat Data Anal 74:95–109

Cherry EC (1953) Some experiments on the recognition of speech, with one and with two ears. J Acoust Soc Am 25(5):975–979

Chohra A, Bahrammirzaee A, Madani K (2013) The impact of irrationality on negotiation strategies with incomplete information. Proceedings of the IADIS International Conference Intelligent Systems and Agents 2013, ISA 2013, Proceedings of the IADIS European Conference on Data Mining 2013, ECDM 2013, pp 3–10

Choma MA, Sarunic MV, Yang C, Izatt JA (2003) Sensitivity advantage of swept source and Fourier domain optical coherence tomography. Opt Express 11(18):2183–2189

Cohen A (2006) Selective Attention. Ency of Cogn Sci. doi:10.1002/0470018860.s00612

Collins GP (2011) Solving the cocktail party problem. Sci Am 304(4):66–67

Dayan P (2014) Rationalizable irrationalities of choice. Top Cogn Sci 6(2):204–228

Ding Y, Dai H, Wang S (2014) Image quality assessment scheme with topographic independent components analysis for sparse feature extraction. Electron Lett 50(7):509–510

Duma M, Twala B, Nelwamondo F, Marwala T (2012) Predictive modeling with missing data using an automatic relevance determination ensemble: a comparative study. Appl Artif Intell 26(10):967–984

Fan J, Wang Y (2014) Fault detection and diagnosis of non-linear non-Gaussian dynamic processes using kernel dynamic independent component analysis. Inform Sciences 259:369–379

Gerschgorin S (1931) Über die Abgrenzung der Eigenwerte einer Matrix. Izv Akad Nauk USSR Otd Fiz-Mat Nauk 6:749–754 [1]

Golub GH, Van Loan CF (1996) Matrix computations. Johns Hopkins University, Baltimore

Gross R, Shi J, Cohn J (2001) Quo vadis face recognition? CMU-RI-TR-01-17, Robotics Institute. Carnegie Mellon University, Pittsburgh

Hafter ER, Xia J, Kalluri S (2013) A naturalistic approach to the cocktail party problem. Adv Exp Med Biol 787:527–534

Hartigan JA (1975) Clustering algorithms. Wiley, New York

Hartigan JA, Wong MA (1979) Algorithm AS 136: a K-means clustering algorithm. J Roy Statist Soc Ser C (Appl Stat) 28(1):100–108

Hawley ML, Litovsky RY, Culling JF (2004) The benefit of binaural hearing in a cocktail party: effect of location and type of interferer. J Acoust Soc Am 115(2):833–843

Haykin S, Chen Z (2005) The cocktail party problem. Neural Comput 17(9):1875–1902

Hess D, Orbe S (2013) Irrationality or efficiency of macroeconomic survey forecasts? Implications from the anchoring bias test. Eur Finan Rev 17(6):2097–2131

Hotelling H (1933) Analysis of a complex of statistical variables into principal components. J Educ Psychol 24:417–441; 498–520

Hyvarinen J, Karhunen E, Oja (2001) Independent component analysis, 1st edn. Wiley, New York

Islam MS, Hannan MA, Basri H, Hussain A, Arebey M (2014) Solid waste bin detection and classification using dynamic time warping and MLP classifier. Waste Manage (Oxford) 34(2):281–290

Jolliffe IT (2002) Principal component analysis, series: Springer Series in statistics, 2nd edn. Springer, NY

Kahneman D (1973) Attention and effort. Prentice-Hall, Englewood Cliffs

Karaman S, Kesler Y, Goksel M, Dogan M, Kayacier A (2014) Rheological and some physicochemical properties of selected hydrocolloids and their interactions with guar gum:

characterization using principal component analysis and viscous synergism index. Int J Food Prop 17(8):1655–1667

Karhunen K (1947) Über lineare Methoden in der Wahrscheinlichkeitsrechnung. Ann Acad Sci Fennicae Ser A I Math-Phys 37:1–79

ldrich J (2006) Eigenvalue, eigenfunction, eigenvector, and related terms. In: Jeff Miller (ed) Earliest known uses of some of the words of mathematics, last retrieved 20 May 2014

Link JV, Lemes ALG, Marquetti I, dos Santos SMB, Bona E (2014) Geographical and genotypic classification of arabica coffee using Fourier transform infrared spectroscopy and radial-basis function networks. Chemometr Intell Lab 135:150–156

Lloyd SP (1982) Least squares quantization in PCM. IEEE Trans Inf Theory 28(2):129–137

Loève M (1978) Probability theory, vol. II, 4th edn. Graduate texts in mathematics 46. Springer, New York

Lu Y, Gao B, Chen P, Charles D, Yu L (2014) Characterisation of organic and conventional sweet basil leaves using chromatographic and flow-injection mass spectrometric (FIMS) fingerprints combined with principal component analysis. Food Chem 154:262–268

MacKay DJC (1991) Bayesian Methods for Adaptive Models. PhD Thesis, California Institute of Technology

MacKay DJC (1992) A practical Bayesian framework for back propagation networks. Neural Comput 4:448–472

Mahalanobis PC (1936) On the generalised distance in statistics. Proc Nat Instit Sci India 2(1):49–55

Malek-Khatabi A, Kompany-Zareh M, Gholami S, Bagheri S (2014) Replacement based non-linear data reduction in radial basis function networks QSAR modeling. Chemomtr Intell Lab 135:157–165

Martinez AR, Benavente R (1998) The AR face database. Technical Report. Computer vision centre technical report. Barcelona, Spain

Marwala T (2001) Fault identification using neural networks and vibration data. Doctor of Philosophy Topic, University of Cambridge

Marwala T (2009) Computational intelligence for missing data imputation, estimation and management: knowledge optimization techniques. IGI Global Publications, information science reference imprint. IGI Global Publications, New York

Marwala T (2012) Condition monitoring using computational intelligence methods. Springer, Heidelberg

Marwala T (2014) Causality, correlation and artificial intelligence for rational decision making. World Scientific Publications, Singapore

Marwala T, Lagazio M (2011) Militarized conflict modeling using computational intelligence techniques. Springer, New York

Metsomaa J, Sarvas J, Ilmoniemi RJ (2014) Multi-trial evoked EEG and independent component analysis. J Neurosci Methods 228:15–26

Moore EH (1920) On the reciprocal of the general algebraic matrix. Bull Amer Math Soc 26(9):394–395

Palmieri F, Fiore U, Castiglione A (2014) A distributed approach to network anomaly detection based on independent component analysis. Concurr Comput Pract Ex 26(5):1113–1129

Pazoki AR, Farokhi F, Pazoki Z (2014) Classification of rice grain varieties using two artificial neural networks (mlp and neuro-fuzzy). J Anim Plant Sci 24(1):336–343

Pearson K (1901) On lines and planes of closest fit to systems of points in space (PDF). Philos Phenomenol 2(11):559–572

Pegram GGS (1983) Spatial relationships in stream flow residuals. Seminar on principal components analysis in the atmospheric and Earth sciences, Pretoria, (CSIR, Pretoria, National Programme for Weather, Climate & Atmosphere Research; CSIR-S-334), pp 134–155

Penrose R (1955) A generalized inverse for matrices. Proc Camb Philos Soc 51:406–413

Preisendorfer RW (1988) Principal component analysis in meteorology and oceanography. Elsevier, Holland

Rad JA, Kazem S, Parand K (2014) Optimal control of a parabolic distributed parameter system via radial basis functions. Commun Nonlinear Sci Numer Simul 19(8):2559–2567

Raol JR (2009) Multi-Sensor data fusion. Theory and practice. CRC Press, Boca Raton

Russ JC (2007) The image processing handbook. CRC Press, Boca Raton

Scharf B (1990) On hearing what you listen for: the effects of attention and expectancy. Can Psychol 31(4):386–387

Schmidt AKD, Römer H (2011) Solutions to the cocktail party problem in insects: selective filters, spatial release from masking and gain control in tropical crickets. Vol. 6. PLoS ONE 6(12) (art. no. e28593)

Schroeder DJ (1999) Astronomical optics, 2nd edn. Academic Press, San Diego, p 433

Setiono R, Liu H (1997) NeuroLinear: from neural networks to oblique decision rules. In: Proceedings: NeuroComputing, vol. 17, pp. 1–24. [View Context]

Shen N (2014) Consumer rationality/irrationality and financial literacy in the credit card market: implications from an integrative review. J Financ Serv Markrting 19(1):29–42

Shutin D, Buchgraber T (2012) Trading approximation quality versus sparsity within incremental automatic relevance determination frameworks. IEEE International Workshop on Machine Learning for Signal Processing, pp 1–6, MLSP art. no. 6349805

Starfield DM, Rubin DM, Marwala T (2006) Near-field artifact reduction using realistic limited-field-of-view coded apertures in planar nuclear medicine imaging. In: Sun I Kim and Tae Suk Sah (eds) Imaging the Future Medicine. Proceedings of the IFMBE vol. 14, pp 3581–3585, Springer, Berlin Heidelberg, ISBN: 978-3-540-36839-7

Starfield DM, Rubin DM, Marwala T (2007) Sampling considerations and resolution enhancement in ideal planar coded aperture nuclear medicine imaging, pp 806-809. 11th Mediterranean Conference on Medical and Biological Engineering June 2007, Ljubljana, Slovenia (IFMBE Proceedings vol. 16) (Paperback) and Computing 2007: MEDICON 2007, 26-30 by Tomaz Jarm, Peter Kramar, Anze Zupanic (eds) Springer, ISBN–10: 3540730435

Stone JV (2004) Independent component analysis: a tutorial introduction. MIT Press, Cambridge

Strang G (1993) Introduction to linear algebra. Wellesley-Cambridge Press, Wellesley

Surajpal D (2007) An independent evaluation of subspace facial recognition algorithms. University of the Witwatersrand Master Thesis

Surajpal D, Marwala T (2007) An independent evaluation of subspace face recognition algorithms. Proceedings of the 18th Annual Pattern Recognition Association of South Africa, ArXiv: 0705.0952

Tan VYF, Févotte C (2013) Automatic relevance determination in nonnegative matrix factorization with the (β)-divergence. IEEE Trans Pattern Anal Mach Intell 35(7):1592–1605. (art. no. 6341758)

Treisman AM (1969) Strategies and models of selective attention. Psychol Rev 76(3):282–299

Varga RS (2002) Matrix iterative Analysis, 2nd edn. (of 1962 Prentice Hall edition), Springer, Englewood Cliffs

Wang XZ, Yang Y, Li R, Mcguinnes C, Adamson J, Megson IL, Donaldson K (2014) Principal component and causal analysis of structural and acute in vitro toxicity data for nanoparticles. Nanotoxicology 8(5):465–476

Wood N, Cowan N (1995) The cocktail party phenomenon revisited: how frequent are attention shifts to one's name in an irrelevant auditory channel? J Exp Psychol Learn Mem Cogn 21(1):255–260

Xu J, Yang J, Shen A, Chen J (2014) A novel ANN-based harmonic extraction method tested with ESN, RNN and MLP in shunt active power filters. Int J Wireless Mobile Comput 7(2):123–131

Yamaguchi N (2012) Variational Bayesian inference with automatic relevance determination for generative topographic mapping. 6th International Conference on Soft Computing and Intelligent Systems, and 13th International Symposium on Advanced Intelligence Systems, SCIS/ISIS 2012, art. no. 6505056, pp 2124–2129

Yan G, Wang X, Li S, Yang J, Xu D (2014) Aberration measurement based on principal component analysis of aerial images of optimized marks. Opt Commun 329:63–68

Yang Y, Nagarajaiah S (2014) Blind identification of damage in time-varying systems using independent component analysis with wavelet transform. Mech Syst Signal Pr 47(1–2):3–20

Zhan X, Dai W, Zhou S (2014) Application of independent component regression in dam deformation analysis. J Inform Comput Sci 11(6):1939–1946

Zhang J, Zhang Z, Zhu G, Chen B, Cheng W, He Z (2014) Multi-unit deflation constraint independent component analysis and its application to source contribution estimation. Jixie Gongcheng Xuebao/J MechE 50(5):57–64+73

Chapter 8
Group Decision Making

8.1 Introduction

There is Tshivenda saying which states that: *Munwe muthihi a u tusi mathuthu* meaning that one finger cannot pick seeds. This old saying is essentially saying that group decision making is always better than an individual decision making. The extension of this to political science is that democracy is always better than one man dictatorship. In a democratic process each voter is given equal weight on the appointment of a government. Therefore, democracy is an example of group decision making.

Wu and Chiclana (2014) applied successfully fuzzy systems, visual information feedback mechanism and attitudinal prioritization technique in group decision making whereas Yue (2014) successfully applied the technique for order preference by similarity to ideal solution and fuzzy systems for group decision making. Wang et al. (2014b) applied successfully multi-criteria group decision-making process which aggregated cloud operators with linguistic information whereas Zhu and Xu (2014) applied successfully fuzzy linear programming technique and additive reciprocal fuzzy preference relations in group decision making.

8.2 Types of Group Decision Making

There are many types of group decision making, some of them effective whereas some of them are not effective, and in this section we describe consensus decision making. For example, one form of group decision making is the consensus based decision making. Consensus decision making seeks to reach a decision by achieving buy-in from all the members of the group. The main disadvantage of consensus decision making is that it does not lead to an optimal decision. The other form of group decision making is by voting. Voting of course does lead to polarization especially for individuals that are defeated.

© Springer International Publishing Switzerland 2014 131
T. Marwala, *Artificial Intelligence Techniques for Rational Decision Making*,
Advanced Information and Knowledge Processing,
DOI 10.1007/978-3-319-11424-8_8

There are four principles for effective decision making: diversity of thought, diversity of information, diversity of capability and capacity to co-operate. Diversity of group is a measure on how uncorrelated decision making drivers of each member of the group is, based on the capacity to think and information. Masisi et al. (2008) compared entropy based methods on measuring structural diversity of an ensemble of 21 classifiers. They successfully applied genetic algorithms to identify the optimal voting based ensemble using the diversity indices as the cost function. This chapter is on the use of artificial intelligence for group decision making and this is described in the next section.

8.3 Artificial Intelligence for Group Decision Making

In artificial intelligence group learning is often called ensemble learning. Ensemble learning is a procedure where multiple models are combined to make a group decision (Sollich and Krogh 1996; Opitz and Maclin 1999; Kuncheva and Whitaker 2003; Polikar 2006; Rokach 2010). Ensemble learning have been proven to, on average, perform better than individual methods and this is described in Appendix B. Zhang et al. (2014) successfully applied ensemble method for medical image classification whereas Pulido et al. (2014) successfully applied ensemble neural networks for prediction of the Mexican Stock Exchange. Wang et al. (2014) applied successfully ensemble of artificial bee colony algorithm as well as Grooms et al. (2014) successfully applied the ensemble Kalman filters in dynamical systems with unresolved turbulence. There are many types of ensembles and these include bagging, stacking, and adaptive boosting.

Bagging is a method which is based on the combination of models fitted to randomly select samples of a training data set to decrease the variance of the prediction model (Breiman 1996a; Samworth 2012; Shinde et al. 2014). In this way bagging attempts to diversify the information that each decision maker has access to. It works by randomly selecting a subset of the training data and training a model and repeating this process. The decision of the group is obtained by combining all decision makers with equal weights to form an ensemble.

Stacking is a technique which exploits the fact that on training a model data can be divided into *training* and *test* data sets. Stacking uses this prior belief by using the performance from the test data to combine the models instead of choosing among them the best performing model by using the test data set (Wolpert 1992; Breiman 1996b; Smyth and Wolpert 1999; Wolpert and Macready 1999; Clarke 2003; Sill et al. 2009).

Boosting is a procedure that incrementally creates an ensemble by training each new model with data that the previously trained model misclassified. Then the ensemble, which is a combination of all trained models, is used for prediction. Adaptive boosting is an extension of boosting to multi-class problems (Schapire 1990; Freund and Schapire 1997; Schapire et al. 1998).

8.3.1 Equally Weighted Ensemble: Support Vector Machines for Land Cover Mapping

In this section an ensemble of support vector machines is used for land cover mapping. The ensemble approach in this figure contains three networks and the output is the equally weighted average of the outputs of these three networks. The ideas presented in this section are the adaptation and the extension of the work by Perrone and Cooper (1993) who introduced the concept of ensemble of networks and it was extended and applied to mechanical systems for the first time by Marwala and Hunt (2000) and this is discussed in detail in Appendix B. The ensembles of networks give results that are more reliable than when using networks separately.

The equally weighted ensemble technique is implemented using three classifiers and these are based on linear support vector machine, radial basis function support vector machines and quadratic support vector machines (SVMs). Support vector machines are supervised learning methods used mostly for classification and are derived from the theory of statistical learning (Vapnik 1995; Cortes and Vapnik 1995). Marwala (2012) applied support vector machines for condition monitoring in structures and Marwala and Lagazio (2011) who applied SVMs in the modelling of militarized interstate conflict. In SVMs a data point is conceptualized as a p-dimensional vector and the objective is to separate such points with a $p-1$-dimensional hyperplane, known as a *linear classifier*. There are many hyperplanes that can be created. Some of these include the one that exhibits the largest separation, also called the *margin*, between the two classes. The selected hyperplane can be chosen so that the distance from it to the nearest data point on both sides is maximized. This is then known as the *maximum-margin hyperplane*. The classification problem can then be stated as estimating a function $f : R^N \rightarrow \{-1,1\}$ dependent on input-output training data, where an independently distributed, unknown probability distribution $P(x, y)$ is chosen such that f can classify unseen (x, y) data (Müller et al. 2001; Habtemariam 2005; Marwala and Lagazio 2011). To apply the linear SVM technique for producing non-linear classifiers, the kernel trick is applied to the maximum-margin hyper-planes (Aizerman et al., 1964; Boser et al., 1992). In this method the dot product is substituted with a non-linear kernel function to fit the maximum-margin hyper-plane in a transformed feature space. Although this dot product transformation may be non-linear, the transformed space may be of high dimensions. For instance, when a linear, radial basis function and quadratic kernels are applied, the resultant feature space is a Hilbert space of infinite dimension. Fei and Bai (2014) applied the SVM for design of turbine blade-tip whereas Garšva and Danenas (2014) successfully applied support vector machines in modelling and control. Yang et al. (2014b) successfully applied SVM for regression and classification with noise while Peng and Xu (2014) successfully applied support vector machine for data classification.

The ensemble of support vector machine was used by Gidudu et al. (2008) and the results are described here using the data from a 2001 Landsat scene of

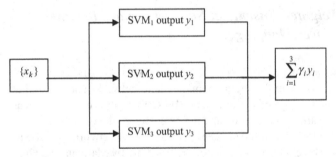

Fig. 8.1 An ensemble of classifiers

Fig. 8.2 Results from linear SVM

Kampala which is the capital of Uganda. These data were input into the SVMs for classification. The ensemble system was built using linear, RBF and a quadratic SVM. Given that the decision boundaries for these classifiers are different, it was assumed they would provide the requisite diversity. The final classification output was determined through a majority vote of the individual classifiers. The ensemble used in this section is shown in Fig. 8.1.

A comparison is made between the derived maps from the individual SVMs and the ensemble system and the results which were reported by Gidudu et al.(2008). Figure 8.2 shows the derived map from the use of linear SVM classifier whereas the derived map from the RBF SVM is in Fig. 8.3 and when quadratic SVM is used the results are in Fig. 8.4. As Gidudu et al.(2008) observed the presence of mixed pixels are observed for linear SVM classifiers whereas the RBF and quadratic SVM classifiers have less mixed pixels. Figure 8.5 shows the result of ensemble system and it can be seen that the ensemble system has greatly enhanced the visual appeal of the linear SVM whereas the results are similar to that of the RBF and quadratic SVMs.

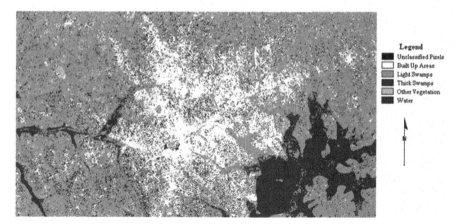

Fig. 8.3 Results from RBF SVM

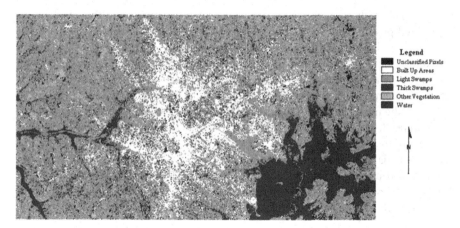

Fig. 8.4 Results from quadratic SVM

8.3.2 Statically Weighted Ensemble: Condition Monitoring in Structures

For this section, the ensemble method which is weighted using the expected error from the test data was applied to identify faults in a population of cylindrical shells. The details of this work can be found in Marwala (2001). Each cylinder was divided into three substructures, and holes of 12 mm diameter were drilled into each substructure. For one cylinder, the first type of fault was a zero-fault scenario, and its identity was [000]. The second type of fault was a one-fault scenario; if it was located in substructure 1, its identity was [100]. The third type of fault was a two-fault scenario, and if the faults were located in substructures 1 and 2, the identity of this case was [110]. The final type of fault was a three-fault scenario, and the identity of this case was [111].

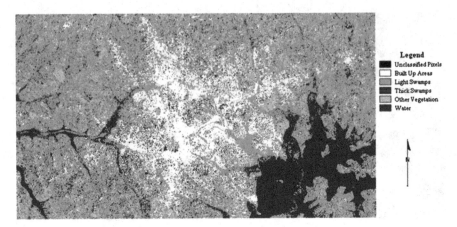

Fig. 8.5 Results from the ensemble

For each fault case, measurements were taken by measuring the acceleration at a fixed position and roving the impulse position. One cylinder gives four fault scenarios and 12 sets of measurements. The structure was vibrated at 19 different locations, nine on the upper ring of the cylinder and ten on the lower ring of the cylinder. Each measurement was taken three times to quantify the repeatability of the measurements. The total number of data points collected was 120.

From the measured data, pseudo-modal energies, modal properties, and wavelet data were identified and used to train the three multi-layer perceptron neural networks and the details of this are in Marwala (2000). The details of modal properties, pseudo-modal energies and wavelet transform are in Appendix A. The WT network was trained using wavelet data. This network had 18 input parameters, 9 hidden units, and 3 output units. The committee was applied using the weighting obtained from the validation data which is described in Appendix B and are as follows:

$$\gamma_1 = \frac{1}{1 + \dfrac{\varepsilon[e_1^2]}{\varepsilon[e_2^2]} + \dfrac{\varepsilon[e_1^2]}{\varepsilon[e_3^2]}} \tag{8.1}$$

$$\gamma_2 = \frac{1}{1 + \dfrac{\varepsilon[e_2^2]}{\varepsilon[e_1^2]} + \dfrac{\varepsilon[e_2^2]}{\varepsilon[e_3^2]}} \tag{8.2}$$

$$\gamma_3 = \frac{1}{1 + \dfrac{\varepsilon[e_3^2]}{\varepsilon[e_1^2]} + \dfrac{\varepsilon[e_3^2]}{\varepsilon[e_2^2]}} \tag{8.3}$$

Table 8.1 Confusion matrix from the classification of fault cases in the test data using the pseudo-modal-energy network

Predicted

		[000]	[100]	[010]	[001]	[110]	[101]	[011]	[111]
	[000]	37	2	0	0	0	0	0	0
	[100]	0	3	0	0	0	0	0	0
	[010]	0	0	3	0	0	0	0	0
Actual	[001]	0	0	0	3	0	0	0	0
	[110]	0	0	0	0	3	0	0	0
	[101]	0	0	0	0	0	3	0	0
	[011]	0	0	0	0	0	0	3	0
	[111]	0	0	0	0	5	1	1	32

Table 8.2 Confusion matrix from the classification of fault cases in the test data using the modal-energy-network

Predicted

		[000]	[100]	[010]	[001]	[110]	[101]	[011]	[111]
	[000]	38	0	0	1	0	0	0	0
	[100]	0	3	0	0	0	0	0	0
	[010]	0	0	3	0	0	0	0	0
Actual	[001]	0	0	0	3	0	0	0	0
	[110]	0	0	0	0	3	0	0	0
	[101]	0	0	0	0	0	3	0	0
	[011]	0	0	0	0	0	0	3	0
	[111]	0	0	0	0	5	2	6	26

when the networks were evaluated using the data not used for training, the results in Table 8.1, 8.2, 8.3, 8.4 were obtained as was done by Marwala (2012). These results indicate that the committee approach gave the best results followed by the pseudo-modal energy network, and then the modal-property network. The wavelet-network performed the worst.

8.3.3 Dynamically Weighted Mixtures of Experts: Platinum Price Prediction

The preceding section describes an effective technique of creating statically weighted ensemble which substantially improved performance. Nevertheless, the data in that example or more significantly the function that produces them is stationary. This section deals with data drawn from a dynamic environment and, therefore, it is not adequate to train the networks and combine them in an

Table 8.3 Confusion matrix from the classification of fault cases in the test data using the -wavelet-network

Predicted

		[000]	[100]	[010]	[001]	[110]	[101]	[011]	[111]
	[000]	35	0	0	1	0	0	0	0
	[100]	0	1	0	0	0	0	0	0
	[010]	2	0	3	0	0	0	0	0
Actual	[001]	0	2	0	2	0	0	0	0
	[110]	0	0	0	0	3	0	0	1
	[101]	0	0	0	1	0	3	0	0
	[011]	0	0	0	0	0	0	3	0
	[111]	1	0	0	0	5	2	6	25

Table 8.4 Confusion matrix from the classification of fault cases in the test data using the committee-network

Predicted

		[000]	[100]	[010]	[001]	[110]	[101]	[011]	[111]
	[000]	38	1	0	0	0	0	0	0
	[100]	0	3	0	0	0	0	0	0
	[010]	0	0	3	0	0	0	0	0
Actual	[001]	0	0	0	3	0	0	0	0
	[110]	0	0	0	0	3	0	0	0
	[101]	0	0	0	0	0	3	0	0
	[011]	0	0	0	0	0	0	3	0
	[111]	0	0	1	0	3	0	1	34

ensemble using static weights derived in Appendix B. The factors which produced the training data are not probably present at the time of testing and, therefore, the ensemble ought to adapt to changing environment. The concept of spatial weightings is extended for time series such that the weightings change as a function of time (Karnick et al. 2008; Lubisky et al. 2008). As described by Lubisky et al. (2008) the weight for each region is updated after each sample and this is conceivable for time series in which, as each sample is received, the correct output for the preceding sample becomes known. Fine-tuning the weights in this way is a potent technique of executing an adaptive model of a system by ensuring that at each time step, the ensemble is updated without retraining each network. If we assume that the features describing a system vary gradually within some bounded space, an ensemble with dynamic weights can maintain its accuracy over time even as the system varies. This model can adapt constantly as long as the conditions of the system were faced in training and consequently effective adaptation is attained, without the cost of retraining the ensemble.

Lubisky et al.(2008) applied this technique to modelling the platinum price which is a particularly hard system to model due to the wide range of factors that impact it. The ensemble was created using the multi-layer perceptron network.

A *neural network* is an information processing method that is stimulated by the way biological nervous systems, like the human brain, process information. It is a computational method aimed at modelling the way in which the brain performs a particular function of interest (Haykin 1999). It is a technique that has been applied to many diverse areas such as estimating dry unit weight of soils (Kolay and Baser 2014), in wastewater treatment (Bagheri et al. 2014), geospatial modelling (Mozumder and Tripathi 2014), forecasting wind power (Yeh et al. 2014) and estimating viscosity (Díaz-Rodríguez et al. 2014). A multi-layer perceptron neural network is a generalized regression model that can model both linear and non-linear data. The building of a neural network consists of the processing units where each has a particular activation level at any point in time; weighted interconnections between a number of processing units that govern how the activation of one unit leads to the input for another unit; an activation rule, which acts on the set of input signals at a processing unit to give a new output signal; and a learning rule that governs how to fine-tune the weights for a given input or output. The architecture of neural processing units and their inter-connections can have a significant influence on the processing capabilities of neural networks. The advantage of the multi-layer perceptron network is the interconnected cross-coupling that occurs between the input variables and the hidden nodes, with the hidden nodes and the output variables.

An ensemble is trained on a full set of training data, 100 samples are adequate to test the ensemble to a large variety of market forces. At each iteration a new random network is produced and trained on a subset of the training data, 20 weeks in this study. Then the network is added to the ensemble and if this decreases the error over the full training set, the network is retained, or else it is thrown out. This technique gives good generalization performance of the ensemble.

The input space was segmented into four regions according to the two features which are deemed most important. The exchange rate and gold prices were applied with the partition along the zero line. The precise choices of areas were not vital, as long as the input space is divided consistently. Each network within the ensemble is allocated a weight in each region corresponding to its performance on samples from that region. The output of the ensemble is the weighted average of the output of each of the experts. In this section the weights of each network is updated. Subsequently when each sample becomes known the weights are recalculated for the 10 previous samples. It is not necessary to retain the past input data, as each network's output will not change. The weight in each region tends exponentially towards one so that if a weight—positive or negative—is not being continuously reinforced, it will lose its significance. This prevents any weight from dominating the ensemble after it has become irrelevant. The performance of an ensemble in predicting the future price of platinum is measured by Eq. 8.4 (Lubinsky et al. 2008):

Table 8.5 The results for different ensembles

Ensemble	4 weeks	10 weeks	20 weeks
Un-weighted	1.14	1.11	1.03
Static weight	0.86	0.97	1.02
Dynamic weight	0.45	0.52	0.74

$$error = \frac{1}{N}\sum_n \left(\frac{y(n)-t_n}{t_n}\right)^2 \qquad (8.4)$$

Here N is the number of data points, y is the prediction and t is the target. Tests were conducted to compare the performance of different models and in each case, 10 ensembles were trained and tested and the average performance is given in Table 8.5 (Lubinsky et al. 2008). The ensembles with constantly updated weights outperform the ensembles which are un-weighted or statically weighted.

8.3.4 Learn++: Wine Recognition

On-line learning is suitable for modelling dynamically time-varying systems where the working conditions vary with time. On-line learning incorporates updated environment resulting from new data. The fundamental feature of on-line learning is incremental learning and it has been researched extensively (Higgins and Goodman 1991; Yamaguchi et al. 1999; Carpenter et al. 1992; Marwala 2012). On-line learning has a disadvantage that it tends to forget the information learned initially in the learning process (McCloskey and Cohen 1989). The on-line learning method used in this chapter was the Learn++ which was proposed by Polikar (Polikar 2000; Polikar et al. 2002; Muhlbaier et al. 2004; Polikar 2006) and (Vilakazi et al. 2006; Vilakazi and Marwala 2007a, b), spectral clustering (Langone et al. 2014), protein classification (Mohamed et al. 2006, 2007), as well as stock market prediction (Lunga and Marwala 2006a, b). Other successful use of incremental learning methods include for anomaly detection (Khreich et al. 2009), missing data estimation (Nelwamondo and Marwala 2007), multi-objective truss design (Pholdee and Bureerat 2014), image segmentation (Yang et al. 2014), online handwriting recognition (Almaksour and Anquetil 2009), reducing the effect of outvoting problem (Erdem et al. 2005), in human robot interaction (Okada et al. 2009), for predicting human and vehicle motion (Vasquez et al. 2009) and in visual learning (Huang et al. 2009). Learn++ is based on adaptive boosting (AdaBoost) and applies multiple classifiers to learn incrementally and this is described in Algorithm 1 proposed by Freund and Schapire (1997).

The algorithm functions by using many classifiers that are weak learners to give a good overall classification. The weakly trained learners are trained on a separate subset of the training data and then the classifiers are combined using a weighted majority vote. The weights for the weighted majority vote are selected using the performance of the classifiers on the entire training dataset.

Each classifier is trained using a training data subset that is sampled in accordance to a desired distribution and using a WeakLearn algorithm. The pre-requisite for the WeakLearn algorithm is that it should give less than 50 % classification rate at the beginning (Polikar et al. 2002). For each database Dk that comprises training sequence, S, where S covers learning data and their equivalent classes, Learn++ begins by resetting the weights, w, in accordance to a desired distribution DT, where T is the number of hypothesis.

Algorithm 1. The Adaptive Boosting algorithm
Input:

- Training data $X = \{\mathbf{x}_1, \mathbf{x}_2, ..., \mathbf{x}_n\}$ with correct labels $\Delta = \{y_1, y_2, ..., y_n\}$
- Weak learn algorithm, known as **Weaklearn**
- Integer T, speciying the number of classifiers $D_1(i) = 1/n; i = 1, ..., n$

For $t = 1, 2, ..., T$;

1. Sample a training subset S_t according to the distribution D_t
2. Train **Weaklearn** with S_t receive hypothesis $h_t : X \rightarrow \Delta$
3. Estimate the error of $h_t : \varepsilon_t = \sum_{i=1}^{n} I[h_t(\mathbf{x}_i) \neq y_i] \cdot D_t(i) = \sum_{t:h_t(\mathbf{x}_i) \neq y_i} D_t(i)$
4. If $\varepsilon_t > \dfrac{1}{2}$ terminate.
5. Estimate the normalized error $\beta_t = \varepsilon_t / (1 - \varepsilon_t) \Rightarrow 0 \leq \beta_t \leq 1$
6. Update the distribution D_t: $D_{t+1}(i) = \dfrac{D_t(i)}{Z_t} \times \begin{cases} \beta_t, & \text{if } h_t(\mathbf{x}_i) y_i \\ 1, & \text{otherwise} \end{cases}$ where Z_t is the

normalization constant so that D_t+1 becomes a proper distribution function.

Test using majority voting given an unlabeled example z as follows:

- Count the total vote from the classifiers $V_j = \sum_{t:h_t(z)} \log(1/\beta_t)$ $j = 1, ..., C$

- Select the class that receives the highest number of votes as the final classification.

The weights are reset to be uniform, in so doing giving equal probability for all examples chosen for the first training subset and the distribution is given by (Polikar et al. 2002):

$$D = 1/m \tag{8.5}$$

Here, m denotes the number of training data in S. The training data are then segmented into training subset TR and testing subset TE to improve ability of the WeakLearn. The distribution is then applied to choose the training subset TR and testing subset TE from S_k. After choosing the training and testing subsets, the WeakLearn algorithm is used. The WeakLearner is trained using subset TR giving the hypothesis, h_t obtained from a WeakLearner and is then tested using both the training and testing subsets to obtain an error (Polikar et al. 2002):

$$\varepsilon_t = \sum_{t:h_t(x_i) \neq y_i} D_t(i) \tag{8.6}$$

The error should be less than 0.5; a normalized error B_t is computed as follows (Polikar et al. 2002):

$$B_t = {}^{\varepsilon_t}/_{1-\varepsilon_t} \tag{8.7}$$

When the error obtained is higher than 0.5, the hypothesis is ignored and we choose a new training and testing subsets according to a distribution D_T and another hypothesis is estimated. All classifiers created are combined using weighted majority voting to give the composite hypothesis, H_t (Polikar et al. 2002):

$$H_t = \arg\max_{y \in Y} \sum_{t:h_t(x)=y} \log({}^1/_{\beta_t}) \tag{8.8}$$

Weighted majority voting provides higher voting weights to a hypothesis that gives lower errors on the training and testing data subsets and the error of the composite hypothesis is thus (Polikar et al. 2002):

$$E_t = \sum_{t:H_t(x_i) \neq y_i} D_t(i) \tag{8.9}$$

When the error is greater than 0.5, the current hypothesis is ignored and the new training and testing data are chosen using the distribution D_T. Or else, if the error is less than 0.5, then the normalized error of the composite hypothesis is calculated as follows (Polikar et al. 2002):

$$B_t = {}^{E_t}/_{1-E_t} \tag{8.10}$$

The error is applied in the distribution update rule, where the weights of the correctly classified case are reduced, accordingly increasing the weights of misclassified instances. This safeguards that the cases that were misclassified by the current hypothesis have a higher probability of being designated for the next training set. The distribution update rule is given by the following equation (Polikar et al. 2002):

$$w_{t+1} = w_t(i) \times B_t^{1-\left[\left|H_t(x_i) \neq y_i\right|\right]} \tag{8.11}$$

After the T hypothesis has been formed for each database, the ultimate hypothesis is calculated by combining the hypotheses using weighted majority voting as described by the following equation (Polikar et al. 2002):

Table 8.6 Data set for whine recognition experiment

Class	C1	C2	C3
S_1	26	31	0
S_2	13	16	32
Validation set	7	8	5
Testing set	13	16	11

$$H_t = \arg\max_{y \in Y} \sum_{k=1}^{K} \sum_{t:H_t(x)=y} \log(1/\beta_t) \qquad (8.12)$$

A method is used to approximate the confidence of the algorithm about its own decision. A majority of hypotheses agreeing on given instances can be interpreted as an indication of confidence on the decision proposed. If it is assumed that a total of T hypotheses are generated in k training sessions for a C-class problem, then for any given example, the final classification class, the total vote class c receives is given by (Muhlbaier et al. 2004):

$$\zeta_c = \sum_{t:h_t(x)=c} \Psi_t \qquad (8.13)$$

here Ψ_t denotes the voting weights of the tth hypothesis h_t. Normalizing the votes received by each class can be performed as follows (Muhlbaier et al. 2004):

$$\lambda_c = \frac{\zeta_c}{\sum_{c=1}^{C} \zeta_c} \qquad (8.14)$$

Here, λ_c can be interpreted as a measure of confidence on a scale of 0–1. A high value of λ_c shows high confidence in the decision and conversely, a low value of λ_c shows low confidence in the decision. It should be noted that the λ_c value does not represent the accuracy of the results, but the confidence of the system in its own decision.

Learn++ was tested on the Wine Recognition Dataset obtained from the University of California, Irvine Repository of Machine Learning (Newman et al. 1998; Hulley and Marwala 2007) The dataset was divided into two training sets S_1 and S_2, a validation set and a testing set respectively. The distribution of the data is shown in Table 8.5 (Hulley and Marwala 2007). Learn++ was simulated 30 times and the optimal number of classifiers were created to classify the data using the validation set with no upper limit. The results and number of classifiers trained per increment of data is shown in Table 8.7 and good results are observed. The generalized performance for the training set S_1 is low because it is being tested on the full testing dataset and not on just the classes that were trained. Thus Learn++ shows good incremental learning ability. (Table 8.6)

Table 8.7 Performance of Learn++ on wine recognition da

Class	C1 (%)	C2 (%)	C3 (%)	C4 (%)	Gen (%)	Std (%)
S_1	96	86	–	70	70	6
S_2	99	87	90	92	92	5

8.3.5 Incremental Learning Using Genetic Algorithm: Optical Character Recognition

The incremental learning approach used in section was proposed by Hulley and Marwala (2007). Unlike the Learn++ approach which uses weak learners to make up a large ensemble of classifiers the Incremental Learning Using Genetic Algorithm (ILUGA) uses strong support vector machines (SVMs) classifiers which are optimized. Each classifier is optimized using genetic algorithm (GA) to identify the optimal separating hyperplane by identifying the best kernel and the best soft margin. Genetic algorithm is an optimization technique which is based on the principles of evolution.

Genetic algorithm has a higher probability of converging to a global optimal solution than a gradient based technique. A genetic algorithm is a population-based, probabilistic method that operates to identify a solution to a problem from a population of possible solutions. It is used to identify approximate solutions to difficult problems through the application of the principles of evolutionary biology to computer science (Michalewicz 1996; Mitchell 1996; Forrest 1996; Vose 1999). It is similar to Darwin's theory of evolution, where members of the population compete to survive and reproduce while the weaker ones die out. Genetic algorithm has been successfully applied to reliability redundancy allocation (Roy et al. 2014), bankruptcy prediction (Gordini 2014), image steganography (Kanan and Nazeri 2014), designing finite impulse response filters (Boudjelaba et al. 2014) as well as optimizing microstructural properties of nanocomposite (Esmaeili and Dashtbayazi 2014).

Each individual is assigned a fitness value according to how well it meets the objective of solving the problem. New and more evolutionary-fit individual solutions are produced during a cycle of generations wherein selection and re-combination operations take place, analogous to how gene transfer applies to current individuals.

The voting weights are then generated by GA using the strong classifiers. ILUGA applies the voting mechanism as used with many ensemble approaches where all the classifiers vote on the class that they predict (Kittler et al. 1998), except that the voting is weighted. Therefore, a weighted majority voting scheme was used. The weights that are sent to the weighted majority voting scheme are individual weights depending on the decision of the classifier. ILUGA allows for as many classifiers as necessary to be trained. The training data for each new classifier that is trained is randomized so that the sections that are used for training and validating, respectively, will always be different giving the classifiers new hypothesis on the data.

The first stage of training for ILUGA is where a strong classifier ensemble is built. The strong classifier ensemble is made up of binary SVM classifiers to classify the multi-class dataset. The training data is randomly separated into three sections namely the training of the classifiers (**Train**), and the two validation sets (**Val1, Val2**). The binary classifiers are trained using the set **Train** which are then evolved using GA to find the optimal parameters for the SVM. The variables to be optimized are the kernel functions (quadratic, radial basis function, polynomial and hyperbolic tangent) and the soft margin. The evolution of the classifiers is done using the validation results from set **Val1** to determine the fitness of each of the chromosomes.

The second stage of training for ILUGA uses the strong binary classifier ensemble that was created in stage one. Each decision of the binary classifiers is assigned with a weight such that each binary classifier has two weights: one for each decision. The weights are then evolved using GA, which are assessed by using the fitness function. The fitness function evaluates the fitness of the chromosomes (weights) by the percentage correctly classified using weighted majority voting on set **Val2**. The optimization of individual weights makes the classification very robust thus eliminating decisions that are incorrectly classified and giving large weightings to correctly classified decisions. This also allows for new classes to be correctly classified as they are seen.

Voting determines the class that is being classified. This is done using the ensemble of classifiers and their corresponding weights. Firstly the number of potential votes for each class is calculated, then the weights for all the classes are divided by their number of potential votes. This gives the new classes that have been introduced the same voting power as the classes that have been seen before. The weights then go into a weighted majority vote where each binary classifier goes to the weighted majority vote and the highest vote is the predicted class system that then goes through stages one and two of the training, to form the new classifiers and weights which are then added to the ensemble. The algorithm used in ILUGA is shown in Algorithm 2 (Hulley and Marwala 2007).

Algorithm 2: Pseudo code for ILUGA
Input

- *Previous binary classifiers*
- *Previous weights*
- *Classes classified by each binary classifier*
- *Number of classifiers to be trained (m)*

Do For *k = 1 to m*

1. *Train strong classifier using set* **Train***, then optimize the parameters using GA with set*

Val1

2. *Optimize the voting weights for each binary decision of the classifier using GA on set*

Table 8.8 Distribution of the OCR data

Class	C0	C1	C2	C3	C4	C5	C6	C7	C8	C9
S_1	250	250	250	0	0	250	250	250	0	0
S_2	150	0	150	250	0	150	0	150	250	0
S_3	0	150	0	150	400	0	150	0	150	400
Test	110	114	111	114	113	115	111	113	110	112

Table 8.9 Performance of ILUGA on the OCR data

Class	C0 (%)	C1 (%)	C2 (%)	C3 (%)	C4 (%)	C5 (%)	C6 (%)	C7 (%)	C8 (%)	C9 (%)	Gen (%)	Std
S_1	100	96	98	–	–	99	99	100	–	–	59	0.2
S_2	100	82	98	94	–	99	84	99	80	–	74	0.9
S_3	92	97	96	92	95	90	99	98	90	83	93	1

Val2

3. *Add the classifiers and weights to the ensemble*
4. *For classification, the number of potential votes for each class is calculated, then the weights for all the classes are divided by their number of potential votes. The voting is done using weighted majority voting to come up with the predicted class*

When ILUGA is going to be incrementally trained, it needs the information of the previous classifiers, the weights and what each of the classifiers is classifying from previous training. The system then goes through stages one and two of the training, to form the new classifiers and weights which are then added to the ensemble.

ILUGA was tested on the optical recognition (OCR) dataset which had ten classes of digits 0–9 and 64 input attributes. The data was split into three sections for the training (S_1, S_2, S_3) and one section for the testing (**Test**). Each of the incremental learning methods were limited to the number of classifiers allowed. ILUGA was allowed to train two multi-class classifiers per dataset. The training dataset was made deliberately challenging to test the ability of the five approaches to learn multiple new classes at once and to retain the information that was previously learnt. The training dataset was set up so that each of the datasets only contain six of the ten classes that are to be trained. Classes 4 and 9 were seen only in the final training set so that the ability of the system to incorporate new classes was fully tested. The distribution of the training and testing datasets is given in Table 8.8. The simulations were run many times to get a generalized average and standard deviation: ILUGA were simulated 30 times and the results are shown in Table 8.9. Each row shows the classification performance per class for the full training dataset. The last two columns show the average generalized performance (**Gen.**) and the standard deviation (**Std.**) of the generalized performance respectively. The generalized performance was done by testing the classifier on the full dataset and not only the classes that were trained for that stage. Thus the generalized performance for sections S_1 and S_2 are low.

Conclusions

In this chapter decision making in groups is studied. In this regard group decision making is effected using artificial intelligence. Four group based decision making techniques were considered and these were ensemble of support vector machines which were successfully applied to land cover mapping, incremental learning using genetic algorithm which was successfully applied to optical character recognition, dynamically weighted mixtures of experts which was successfully applied to platinum price prediction as well as Learn++ which was successfully applied to wine recognition.

References

Aizerman M, Braverman E, Rozonoer L (1964) Theoretical foundations of the potential function method in pattern recognition learning. Autom Rem Contr 25:821–837

Almaksour A, Anquetil E (2009) Fast incremental learning strategy driven by confusion reject for online handwriting recognition. Proceedings of the international conference on document analysis and recognition, 81–85

Bagheri M, Mirbagheri SA, Ehteshami M, Bagheri Z (2014) Modeling of a sequencing batch reactor treating municipal wastewater using multi-layer perceptron and radial basis function artificial neural networks, Process Safety and Environmental Protection. Available online 30

Boser BE, Guyon IM, Vapnik VN (1992) A training algorithm for optimal margin classifiers. In: Haussler D (ed) 5th Annual ACM workshop on COLT. ACM Press, Pittsburgh

Boudjelaba K, Ros F, Chikouche D (2014) An efficient hybrid genetic algorithm to design finite impulse response filters. Expert Syst Appli 41(13):5917–5937

Breiman L (1996a) Stacked regression. Mach Learn 24:49–64

Breiman L (1996b) Bagging predictors. Mach Learn 24 (2):123–140

Carpenter GA, Grossberg S, Marhuzon N, Reynolds JH, Rosen DB (1992) ARTMAP: a neural network architecture for incremental learning supervised learning of analog multidimensional maps. IEEE Trans on Neural Nets 3:698–713

Clarke B (2003) Bayes model averaging and stacking when model approximation error cannot be ignored. Mach Learn Res 4:683–712

Cortes C, Vapnik V (1995) Support-vector networks. Mach Learn 20(3):273

Díaz-Rodríguez P, Cancilla JC, Matute G, Torrecilla JS (2014) Viscosity estimation of binary mixtures of ionic liquids through a multi-layer perceptron model, Journal of Industrial and Engineering Chemistry, Available online 11 June 2014

Erdem Z, Polikar R, Gurgen F, Yumusak N (2005) Reducing the effect of out-voting problem in ensemble based incremental support vector machines. Lecture Notes in Comp Sci 3697:607–612

Esmaeili R, Dashtbayazi MR (2014) Modeling and optimization for microstructural properties of Al/SiC nanocomposite by artificial neural network and genetic algorithm. Expert Syst Appl 41(13):5817–5831

Fei C-W, Bai G-C (2014) Distributed collaborative probabilistic design for turbine blade-tip radial running clearance using support vector machine of regression. Mech Syst Signal Process 49(1–2)196–208

Forrest S (1996) Genetic algorithms. ACM Comput Surv 28:77–80

Freund Y, Schapire RE (1997) Decision-theoretic generalization of on-line learning and an application to boosting. J Comp Syst Sci 55:119–139

Garšva G, Danenas P (2014) Particle swarm optimization for linear support vector machines based classifier selection. Nonlinear Anal:Model Control 19(1):26–42

Gidudu G, Hulley G, Marwala T (2008) An SVM multiclassifier approach to land cover mapping. ASPRS 2008 Annual Conference Portland, Oregon â¦ April 28–May 2, 2008

Gordini N (2014) A genetic algorithm approach for SMEs bankruptcy prediction: empirical evidence from Italy. Expert Syst Appl 41(14):6433–6445

Grooms I, Lee Y, Majda AJ (2014) Ensemble Kalman filters for dynamical systems with unresolved turbulence. J Computational Physics 273:435–452

Habtemariam E, Marwala T, Lagazio M (2005) Artificial intelligence for conflict management. Proceedings of the IEEE international joint conference on neural networks, Montreal, Canada, pp 2583–2588.

Haykin S (1999) Neural networks. Prentice-Hall, New Jersey

Higgins CH, Goodman RM (1991) Incremental learning for rule based neural network. In: Proceeding of the International Joint Conference on Neural Networks, 875–880

Huang D, Yi Z, Pu X (2009) A new incremental PCA algorithm with application to visual learning and recognition. Neural Process Lett 30:171–185

Hulley G, Marwala T (2007) Genetic algorithm based incremental learning for optimal weight and classifier selection. Proceedings of the AIP Conference, 258–267

Kanan HR, Nazeri B (2014) A novel image steganography scheme with high embedding capacity and tunable visual image quality based on a genetic algorithm. Expert Syst Appl 41(14):6123–6130

Karnick M, Ahiskali M, Muhlbaier MD, Polikar R (2008) Learning concept drift in non-stationary environments using an ensemble of classifiers based approach. International Joint Conference on Neural Networks

Khreich W, Granger E, Miri A, Sabourin RA (2009) A comparison of techniques for on-line incremental learning of HMM parameters in anomaly detection. Proceedings of the IEEE Symposium on Computational Intelligence for Security and Defense Applications, 1–8

Kittler J, Hatef M, Duin R, Matas J (1998) On combining classifiers. IEEE Trans Pattern Anal Mach Intell.20:226–238

Kolay E, Baser T (2014) Estimating of the dry unit weight of compacted soils using general linear model and multi-layer perceptron neural networks. Appl Soft Comput 18:223–231

Kuncheva L, Whitaker C (2003) Measures of diversity in classifier ensembles. Mach Learn 51:181–207

Langone R, Mauricio Agudelo O, De Moor B, Suykens JAK (2014) Incremental kernel spectral clustering for online learning of non-stationary data. Neurocomputing 139:246–260

Lubinsky B, Genc B, Marwala T (2008) Prediction of platinum prices using dynamically weighted mixture of experts. arXiv:0812.2785

Lunga D, Marwala T (2006a) Time series analysis using fractal theory and online ensemble classifiers. Lect Notes in Comp Sci 4304:312–321

Lunga D, Marwala T (2006b) Online forecasting of stock market movement direction using the improved incremental algorithm. Lect Notes in Comp Sci 4234:440–449

Marwala T (2000) On damage identification using a committee of neural networks. American Society of Civil Engineers. J Eng Mech 126:pp 43–50

Marwala T (2001) Fault identification using neural networks and vibration data. Doctor of Philosophy Topic, University of Cambridge

Marwala T (2012) Condition monitoring using computational intelligence methods. Springer-Verlag, Heidelberg

Marwala T, Hunt HEM (2000) Probabilistic fault identification using vibration data and neural networks. Proceedings of SPIE—The International Society for Optical Engineering, 4062, ISSN 0277-786X

Marwala T, Lagazio M (2011) Militarized conflict modeling using computational intelligence techniques. Springer, Heidelberg

Masisi L, Nelwamondo V, Marwala T (2008) The use of entropy to measure structural diversity. arXiv:0810.3525

McCloskey M, Cohen N (1989) Catastrophic interference connectionist networks: the sequential learning problem. The Psychol of Learn and Motivat 24:109–164

Michalewicz Z (1996) Genetic Algorithms + Data Structures = Evolution Programs. New York, Springer

Mitchell M (1996) An Introduction to Genetic Algorithms. MIT Press, Cambridge

Mohamed S, Rubin D, Marwala T (2006) Multi-class protein sequence classification using fuzzy ARTMAP. Proceeding of the IEEE International Conference on System, Man and Cybernetics, 1676–1681

Mohamed S, Rubin D, Marwala T (2007) Incremental learning for classification of protein sequences. Proceeding of the IEEE International Joint Conference on Neural Networks, 19–24

Mozumder C, Tripathi NK (2014) Geospatial scenario based modelling of urban and agricultural intrusions in Ramsar wetland Deepor Beel in Northeast India using a multi-layer perceptron neural network. Intl J of Appl Earth Obs and Geoinform 32:92–104

Muhlbaier M, Topalis A, Polikar R (2004) Learn++.MT: a new approach to incremental learning. In: Proceedings of the 5th International Workshop on Multiple Classifier Systems,52–61

Müller KR, Mika S, Ratsch G, Tsuda K, Scholkopf B (2001) An Introduction to kernel-based learning algorithms. IEEE Trans on Neur Nets 12:181–201

Nelwamondo FV, Marwala T (2007) Handling missing data from heteroskedastic and nonstationary data. Lect Notes in Comp Sci 4491:1293–1302

Newman CBDJ, Hettich S, Merz C (1998) UCI repository of machine learning databases. University of California, Irvine, Dept. of Information and Computer Sciences. http://www.ics.uci.edu/_mlearn/MLRepository.html. Accessed 1 Feb 2014

Okada S, Kobayashi Y, Ishibashi S, Nishida T (2009) Incremental learning of gestures for human-robot interaction. AI and Soc 25:155–168

Opitz D, Maclin R (1999) Popular ensemble methods: an empirical study. J Artif Intell Res 11:169–198

Peng X, Xu D (2014) Structural regularized projection twin support vector machine for data classification. Inf Sci 279:416–432

Perrone MP, Copper LN (1993) When networks disagree: ensemble methods for hybrid neural network. In: Mammone RJ (ed) Neural networks for speech and image processing. Chapman-Hall, London, pp 126–142

Pholdee N, Bureerat S (2014) Hybrid real-code population-based incremental learning and approximate gradients for multi-objective truss design. Eng Optim 46(8):1032–1051

Polikar R (2000) Algorithms for enhancing pattern separability, feature selection and incremental learning with applications to gas sensing electronic noise systems. PhD thesis, Iowa State University, Ames

Polikar R (2006) Ensemble based systems in decision making. IEEE Circuits and Syst Mag 6: 21–45

Polikar R, Byorick J, Krause S, Marino A, Moreton M (2002) Learn++: a classifier independent incremental learning algorithm for supervised neural network. Proceeding of International Joint Conference on Neural Networks, 1742–1747

Pulido M, Melin P, Castillo O (2014) Particle swarm optimization of ensemble neural networks with fuzzy aggregation for time series prediction of the Mexican Stock Exchange. Inf Sci 280, 188–204

Rokach L (2010) Ensemble-based classifiers. Artif Intell Rev 33(1–2):1–39

Roy P, Mahapatra BS, Mahapatra GS, Roy PK (2014) Entropy based region reducing genetic algorithm for reliability redundancy allocation in interval environment. Expert Syst Appl 41(14):6147–6160

Samworth RJ (2012) Optimal weighted nearest neighbour classifiers. Ann Stat 40(5):2733–2763

Schapire RE (1990) The strength of weak learnability. Mach Learn 5:197–227

Schapire RE, Freund Y, Bartlett P, Lee WS (1998) Boosting the margin: a new explanation for the effectiveness of voting methods. Ann Stat 26:51–1686

Shinde A, Sahu A, Apley D, Runger G (2014) Preimages for variation patterns from kernel PCA and bagging. IIE Trans 46(5)

Sill J, Takacs G, Mackey L, Lin D (2009) Feature-weighted linear stacking arXiv:0911.0460

Smyth P, Wolpert DH (1999) Linearly combining density estimators via stacking. Mach Learn J 36:pp 59–83

Sollich P, Krogh A (1996) Learning with ensembles: how overfitting can be useful. Adv Neural Inf Process Syst 8:190–196

Vapnik, VN (1995) The nature of statistical learning theory Springer, New York

Vasquez D, Fraichard T, Laugier C (2009) Growing hidden Markov models: an incremental tool for learning and predicting human and vehicle motion. Intl J of Robot Res 28:1486–1506

Vilakazi CB, Marwala T (2007a) Incremental learning and its application to bushing condition monitoring. Lect Notes in Comp Sci 4491:1237–1246

Vilakazi CB, Marwala T (2007b) Online incremental learning for high voltage bushing condition monitoring. Proceeding of International Joint Conference on Neural Networks, 2521–2526

Vilakazi CB, Marwala T, Mautla R, Moloto E (2006) Online bushing condition monitoring using computational intelligence. WSEAS Trans Power Syst 1:280–287

Vose MD (1999) The simple genetic algorithm: foundations and theory. MIT Press, Cambridge

Wang H, Wu Z, Rahnamayan S, Sun H, Liu Y, Pan, J-S (2014a) Multi-strategy ensemble artificial bee colony algorithm. Inf Sci 279:587–603

Wang J-Q, Peng L, Zhang H-Y, Chen X-H (2014b) Method of multi-criteria group decision-making based on cloud aggregation operators with linguistic information. Inf Sci 274:177–191

Wolpert D (1992) Stacked generalization. Neural Netw 5(2):241–259

Wolpert DH, Macready WG (1999) An efficient method to estimate bagging's generalization error. Mach Learn J 35:41–55

Wu J, Chiclana F (2014) Visual information feedback mechanism and attitudinal prioritisation method for group decision making with triangular fuzzy complementary preference relations. Inf Sci 279:716–734

Yamaguchi K, Yamaguchi N, Ishii N (1999) Incremental learning method with retrieving of interfered patterns. IEEE Trans on Neural Nets 10:1351–1365

Yang S, Lv Y, Ren Y, Yang L, Jiao L (2014a) Unsupervised images segmentation via incremental dictionary learning based sparse representation. Inf Sci 269:48–59

Yang X, Tan L, He L (2014b) A robust least squares support vector machine for regression and classification with noise. Neurocomputing 140:41–52

Yeh W-C, Yeh Y-M, Chang P-C, Ke Y-C, Chung V (2014) Forecasting wind power in the Mai Liao Wind Farm based on the multi-layer perceptron artificial neural network model with improved simplified swarm optimization. Intl J Electr Power & Energy Syst 55: 741–748

Yue Z (2014) TOPSIS-based group decision-making methodology in intuitionistic fuzzy setting. Inf Sci 277:141–153

Zhang Y, Zhang B, Coenen F, Xiao J, Lu W (2014) One-class kernel subspace ensemble for medical image classification. Eurasip J Adv Signal Process 2014(1):17

Zhu B, Xu Z (2014) A fuzzy linear programming method for group decision making with additive reciprocal fuzzy preference relations. Fuzzy Sets Syst 246:19–33

Chapter 9
Conclusion

9.1 Introduction

This book was on rational decision making with the aide of artificial intelligence. The classical definition of a rational agent is an agent which acts to maximize its utility (Osborne and Rubinstein 2001; Russell and Norvig 2003). Utility is a difficult concept to grasp. It is classically defined as the ability of an object to satisfy needs. Much has been written about utility and its derivative expected utility. One definition of utility which dominates the economics field is that it is a representation of the preference of some goods or services (Berger 1985; Ingersoll 1987; Castagnoli and LiCalzi 1996; Bordley and LiCalzi 2000). Samuelson attempted to quantify utility as a measure of the willingness of the people to pay for a particular good (Samuelson 1938). In this book we view utility as a measure of the value of some goods less the cost associated with the acquisition of such goods. This simply means that we define utility as the value that is derived from a good minus the cost of that good. Thus if the good is quite valuable but is also equally expensive then its utility is zero because its value and cost balance out.

The other aspect of this book is artificial intelligence which is simply a computational paradigm inspired by the natural world and all its inhabitants to build systems that are deemed intelligent (Embrechts et al. 2014). In this book there are many different strands of artificial intelligence methods which have been considered and these include support vector machines, rough sets, fuzzy system, neural networks, particle swarm optimization, genetic algorithm and simulated annealing (Zadeh 1965; Rumelhart and McClelland 1986; Nguyen 1990; Fogel et al. 1966). Artificial intelligence techniques have been widely used to solve many complex problems such as in photovoltaic design (Maleki and Askarzadeh 2014) and modelling river bed material discharge (Roushangar et al. 2014). This book applies artificial intelligence for rational decision making and defines rational decision making as consisting of three core elements: (1) Using relevant information to make a decision; (2) Using the correct logic to arrive at a decision; and (3) Optimizing the whole decision process including maximizing the utility.

© Springer International Publishing Switzerland 2014
T. Marwala, *Artificial Intelligence Techniques for Rational Decision Making,*
Advanced Information and Knowledge Processing,
DOI 10.1007/978-3-319-11424-8_9

Conclusions and Further Work

This book proposes that rational decision making is executed by using two primary drivers and these are causality and correlation. In fact embedded in artificial intelligence methods for prediction which is the basis of rational decision making using artificial intelligence are the correlation and causal machines. This book defines a causal function as the function which maps the input to the output where there is a flow of information from the cause to the effect. In philosophy this is what is termed as the transmission theory of causality. The importance of causality on understanding many vital systems has been studied in fields such as child development (Kegel and Bus 2014), wind turbines (Tippmann and Scalea 2014), heart studies (Seiler et al. 2013) and in philosophy (Chaigneau and Puebla 2013).

This book formulated causal function within the context of rational decision making. The causal function was implemented using rough sets to maximize the attainment of an optimal decision. The rough sets which is a type of artificial intelligence was used to identify the causal relationship between the militarized interstate dispute (MID) variables (causes) and conflict status (effects) such that whenever the MID variables were given the probability of conflict was then estimated. Rough sets has modelled diverse problems successfully such as rural sustainability development (Boggia et al. 2014), in web search (Yahyaoui et al. 2014), fault classification of gas turbines (Li et al. 2014), customer segmentation (Dhandayudam and Krishnamurthi 2014) and hepatitis (Srimathi and Sairam 2014). As a way forward other artificial intelligence approaches should be explored to classify them in terms of effectiveness on the construction of a causal function given a particular application and hopefully a generalized conclusion can then be drawn.

Another important area that is studied in this book is the concept of correlation function which is a function which maps the input to the output where there is no flow of information from the cause to the effect. The concept of correlation has been applied in many areas such as constructing a function generator (Li et al. 2014), to study the mechanism of cell aggregation (Agnew et al. 2014), in crystal studies (Jaiswal et al. 2014) and in damage detection (Ni et al. 2014). The correlation function is implemented using support vector machine and succesfully applied to identify the correlation relationship between the electroencephalogram (EEG) signal and eliptic activity of patients. Support vector machines have been used successfully in the past to model EEG signal by Yin and Zhang (2014), Kumar et al. (2014), Yu et al. (2013), Panavaranan and Wongsawat (2013) as well as Bhattacharyya et al. (2013). For future work other artificial intelligence techniques should be applied in other complex examples to model a correlation function.

This book also introduced the concept of missing data estimation as a mechanism for rational decision making. This assumed that there is a fixed topological characteristic between the variables required to make a rational decision and the actual rational decision. This technique was successfully operationalized using an autoassociative multi-layer perceptron network trained using the scaled conjugate method (Chen 2010; Mistry et al. 2008; Duma et al. 2012) and the missing data were estimated using genetic algorithm (Aydilek and Arslan 2013; Azadeh et al. 2013; Canessa et al. 2012; Devi Priya and Kuppuswami 2012; Hengpraphrom et al. 2011). This technique was successful used to predict HIV status of a subject given the demographic characteristics. For the future other missing data estimation techniques should be used for rational decision making.

This book introduced the concept of rational countefactuals which is an idea of identifying a counterfactual from the factual and knowledge of the laws that govern the relationships between the antecedent and the consequent, that maximizes the attainment of the desired consequent (Hausman and Woutersen 2014; Scholl and Sassenberg 2014; Vaidman 2014; Van Hoeck et al. 2014; Wang and Ma 2014). This is primarily intended to either avoid previous mistakes or reinforce previous successes. In this chapter in order to build rational counterfactuals neuro-fuzzy model (Airaksinen 2001, 2004) and genetic algorithm were applied (Martin and Quinn 1996; De Faria and Phelps 2011). The theory of rational counterfactuals was applied to identify the antecedent that assure the attainment of peaceful outcome in a conflict situation. For the future the concept of rational counterfactuals should be applied to other complex problems such as in economics and engineering sciences.

This book advanced the theory of flexibly bounded rationality which is an extension of Herbert Simon's theory of bounded rationality where rationality is bounded because of inadequate information to make a decision, limited processing capability and limited brain power (Ding et al. 2014a, b; Shi et al. 2014). In flexibly bounded rationality inadequate information is variable because of advances in missing data estimation techniques, processing power is variable because of Moore's Law where computational limitations are advanced continuously and limited brain power which is more and more being replaced with artificial intelligence techniques (Frenzel 2014). The multi-layer perceptron network and particle swarm optimization were applied to implement the theory of flexibly bounded rationality in the problem of interstate conflict. For the future work, this technique should be tested in other complex problems and using other artificial intelligence techniques.

This book dealt with the concept of using relevant information as a basis of rational decision making. It proposed three methods for making rational decisions by either marginalizing irrelevant information or not using irrelevant information. In this regard four techniques were considered and these were marginalization of irrationality approach, automatic relevance determination (Smyrnakis and Evans 2007; Fu and Browne 2007), principal component analysis (Evans and Kennedy 2014; Yan et. al. 2014) and independent component analysis (Zhan et al. 2014; Chen and Yu 2014). These techniques were applied to condition monitoring, credit scoring, interstate conflict and face recognition. For the future these techniques should be applied to other more complex problems.

This book considered the concept of group decision making and how artificial intelligence is used to facilitate decision making in a group (Bolón-Canedo et al. 2014; Hailat et al. 2014). Four group based decision making techniques were considered and these were ensemble of support vector machines which were applied to land cover mapping, condition monitoring, incremental learning using genetic algorithm which was applied to optical character recognition, dynamically weighted mixtures of experts which were applied to platinum price prediction as well as the Learn ++ which was applied to wine recognition. For the future, other forms of ensemble such as products of experts should be considered for further studies.

References

Agnew DJG, Green JEF, Brown TM, Simpson MJ, Binder BJ (2014) Distinguishing between mechanisms of cell aggregation using pair-correlation functions. J Theor Biol 352:16–23

Airaksinen T (2001) Counterfactuals and other philosophical challenges to machine intelligence: a fuzzy view. Annual Conference of the North American Fuzzy Information Processing Society—NAFIPS, vol 5, pp 2930–2934

Airaksinen T (2004) What a machine should know about philosophical problems? Soft Comput 8(10):650–656

Aydilek IB, Arslan A (2013) A hybrid method for imputation of missing values using optimized fuzzy c-means with support vector regression and a genetic algorithm. Inf Sci 233:25–35

Azadeh A, Asadzadeh SM, Jafari-Marandi R, Nazari-Shirkouhi S, Baharian Khoshkhou G, Talebi S, Naghavi A (2013) Optimum estimation of missing values in randomized complete block design by genetic algorithm. Knowl-Based Syst 37:37–47

Berger JO (1985) Utility and loss. Statistical decision theory and Bayesian analysis, 2nd edn. Springer, Berlin

Bhattacharyya S, Rakshit P, Konar A, Tibarewala DN, Janarthanan R (2013) Feature selection of motor imagery EEG signals using firefly temporal difference Q-learning and support vector machine. Lecture Notes in Computer Science (including subseries Lecture Notes in Artificial Intelligence and Lecture Notes in Bioinformatics), 8298 LNCS (Part 2), pp 534–545

Boggia A, Rocchi L, Paolotti L, Musotti F, Greco S (2014) Assessing rural sustainable development potentialities using a dominance-based rough set approach. J Environ Manage 144:160–167

Bolón-Canedo V, Sánchez-Maroño N, Alonso-Betanzos A (2014) Data classification using an ensemble of filters. Neurocomputing 135:13–20

Bordley R, LiCalzi M (2000) Decision analysis with targets instead of utilities. Decis Econ Finance 23:53–74

Canessa E, Vera S, Allende H (2012) A new method for estimating missing values for a genetic algorithm used in robust design. Eng Optim 44(7):787–800

Castagnoli E, LiCalzi M (1996) Expected utility theory without utility. Theory Decis 41:281–301

Chaigneau SE, Puebla G (2013) The proper function of artifacts: intentions, conventions and causal inferences. Rev Phil Psychol 4(3):391–406

Chen J, Yu J (2014) Independent component analysis mixture model based dissimilarity method for performance monitoring of non-Gaussian dynamic processes with shifting operating conditions. Ind Eng Chem Res 53(13):5055–5066

Chen M-H (2010) Pattern recognition of business failure by autoassociative neural networks in considering the missing values. ICS 2010—International Computer Symposium, art. no. 5685421, pp 711–715

De Faria LG, Phelps S (2011) An investigation of the consequences of basel III using an agent-based model. ACM International Conference Proceeding Series, art. no. 2378131

Devi Priya R, Kuppuswami S (2012) A genetic algorithm based approach for imputing missing discrete attribute values in databases. WSEAS T Inf Sci Appl 9(6):169–178

Dhandayudam P, Krishnamurthi I (2014) A rough set approach for customer segmentation. Data Sci J 13:1–11

Ding Z, Li Q, Ge D, Jiang S (2014a) Research on dynamics in a resource extraction game with bounded rationality. Appl Math Comput 236:628–634

Ding Z, Zhu X, Jiang S (2014b) Dynamical Cournot game with bounded rationality and time delay for marginal profit. Math Comput Simul 100:1–12

Duma M, Twala B, Marwala T, Nelwamondo FV (2012) Classification with missing data using multi-layered artificial immune systems. 2012 IEEE Congress on Evolutionary Computation, CEC 2012, art. no. 6256420

Embrechts MJ, Rossi F, Schleif F-M, Lee JA (2014) Advances in artificial neural networks, machine learning, and computational intelligence (ESANN 2013). Neurocomputing 141:1–2

Evans M, Kennedy J (2014) Integration of adaptive neuro fuzzy inference systems and principal component analysis for the control of tertiary scale formation on tinplate at a hot mill. Expert Systems Appl 41(15):6662–6675

Fogel LJ, Owens AJ, Walsh MJ (1966) Artificial intelligence through simulated evolution. Wiley, New York

Frenzel L (2014) Is Moore's law really over for good? Electronic Des 62(3)

Fu Y, Browne A (2007) Using ensembles of neural networks to improve automatic relevance determination. IEEE International Conference on Neural Networks—Conference Proceedings, art. no. 4371195, pp 1590–1594

Hailat E, Russo V, Rushaidat K, Mick J, Schwiebert L, Potoff J (2014) Parallel Monte Carlo simulation in the canonical ensemble on the graphics processing unit. Int J Parallel Emerg Distrib Syst 29(4):379–400

Hausman J, Woutersen T (2014) Estimating the derivative function and counterfactuals in duration models with heterogeneity. Econometric Rev 33(5–6):472–496

Hengpraphrom K, Wlchian SN, Meesad P (2011) Missing value imputation using genetic algorithm. ICIC Express Letters 5(2):355–360

Ingersoll JE Jr (1987) Theory of financial decision making. Rowman and Littlefield, Totowa

Jaiswal A, Bharadwaj AS, Singh Y (2014) Communication: integral equation theory for pair correlation functions in a crystal. J Chem Phys 140(21). art. no. 211103

Kegel CAT, Bus AG (2014) Evidence for causal relations between executive functions and alphabetic skills based on longitudinal data. Infant and. Child Dev 23(1):22–35

Kumar D, Tripathy RK, Acharya A (2014) Least square support vector machine based multiclass classification of EEG signals. WSEAS T Signal Process 10(1):86–94

Li F, Chang C-H, Basu A, Siek L (2014) A 0.7 V low-power fully programmable Gaussian function generator for brain-inspired Gaussian correlation associative memory. Neurocomputing 138:69–77

Li Y-D, Li H-W, Zhang B-C, Yang J, Liu H-Y, Zhang J (2014) Fault diagnosis of gas turbine generator set by combination of rough sets and neural network. Power Syst Prot Control 42(8):90–94. (Dianli Xitong Baohu yu Kongzhi)

Maleki A, Askarzadeh A (2014) Comparative study of artificial intelligence techniques for sizing of a hydrogen-based stand-alone photovoltaic/wind hybrid system. Int J Hydrogen Energy 39(19):9973–9984

Martin AD, Quinn KM (1996) Using computational methods to perform counterfactual analyses of formal theories. Ration Soc 8(3):295–323

Mistry J, Nelwamondo FV, Marwala T (2008) Using principal component analysis and autoassociative Neural Networks to estimate missing data in a database. WMSCI 2008—The 12th World Multi-Conference on Systemics, Cybernetics and Informatics, Jointly with the 14th International Conference on Information Systems Analysis and Synthesis, ISAS 2008—Proc. 5, pp 24–29

Nguyen DH (1990) Neural networks for self-learning control systems. Control Syst Mag IEEE 10(3):18–23

Ni P, Xia Y, Law S-S, Zhu S (2014) Structural damage detection using auto/cross-correlation functions under multiple unknown excitations. Int J Struct Stab Dynam 14(5). (Article in Press)

Osborne M, Rubinstein A (2001) A course in game theory. MIT, Cambridge

Panavaranan P, Wongsawat Y (2013) EEG-based pain estimation via fuzzy logic and polynomial kernel support vector machine. BMEiCON 2013—6th Biomedical Engineering International Conference, art. no. 6687668

Roushangar K, Mehrabani FV, Shiri J (2014) Modeling river total bed material load discharge using artificial intelligence approaches (based on conceptual inputs). J Hydrol 514:114–122

Rumelhart DE, McClelland J (1986) Parallel distributed processing: explorations in the micro-structure of cognition. MIT, Cambridge

Russell SJ, Norvig P (2003) Artificial intelligence: a modern approach, 2nd edn. Prentice Hall, New Jersey

Samuelson PA (1938) A note on measurement of utility. Rev Econ Stud 4(2):155–161

Scholl A, Sassenberg K (2014) Where could we stand if I had...? How social power impacts coun-terfactual thinking after failure. J Exp Soc Psychol 53:51–31

Seiler C, Engler R, Berner L, Stoller M, Meier P, Steck H, Traupe T (2013) Prognostic relevance of coronary collateral function: confounded or causal relationship? Heart 99(19):1408–1414

Shi L, Le Y, Sheng Z (2014) Analysis of price Stackelberg duopoly game with bounded rationality. Discrete Dyn Nat Soc 2014 art. no. 428568. doi:10.1155/2014/428568

Smyrnakis MG, Evans DJ (2007) Classifying ischemic events using a Bayesian inference multi-layer perceptron and input variable evaluation using automatic relevance determination. Com-put Cardiol 34:305–308 art. no. 4745482

Srimathi S, Sairam N (2014) A soft computing system to investigate hepatitis using rough set re-ducts classified by feed forward neural networks. Int J Appl Eng Res 9(10):1265–1278

Tippmann JD, Di Scalea FL (2014) Experiments on a wind turbine blade testing an indication for damage using the causal and anti-causal Green's function reconstructed from a diffuse field. Proceedings of SPIE—The International Society for Optical Engineering, 9064, art. no. 90641I

Vaidman L (2014) Comment on protocol for direct counterfactual quantum communication. Phys Rev Lett 112(20) art. no. 208901

Van Hoeck N, Begtas E, Steen J, Kestemont J, Vandekerckhove M, Van Overwalle F (2014) False belief and counterfactual reasoning in a social environment. Neuroimage 90:315–325

Wang L, Ma W-F (2014) Comparative syllogism and counterfactual knowledge. Synthese 191(6):1327–1348

Yahyaoui H, Almulla M, Own HS (2014) A novel non-functional matchmaking approach between fuzzy user queries and real world web services based on rough sets. Future Gener Comput Syst 35:27–38

Yan G, Wang X, Li S, Yang J, Xu D (2014) Aberration measurement based on principal component analysis of aerial images of optimized marks. Opt Commun 329:63–68

Yin Z, Zhang J (2014) Identification of temporal variations in mental workload using locally-linear-embedding-based EEG feature reduction and support-vector-machine-based clustering and classification techniques. Comput Method Programs Biomed 115(3):119–134

Yu S, Li P, Lin H, Rohani E, Choi G, Shao B, Wang Q (2013) Support vector machine based detec-tion of drowsiness using minimum EEG feature. Proceedings—SocialCom/PASSAT/BigData/EconCom/BioMedCom 2013, art. no. 6693421, pp 827–835

Zadeh LA (1965) Fuzzy sets. Inf Control 8(3):338–353

Zhan X, Dai W, Zhou S (2014) Application of independent component regression in dam deforma-tion analysis. J Inf Comput Sci 11(6):1939–1946

Appendices

A.1 Fourier Transform, Wavelet Transform, Modal Properties and Pseudo-Modal Energies

A.1.1 Fourier Transform

The Fourier Transform is a mathematical technique for transforming information in the time domain into the frequency domain. Inversely, the inverse Fourier transform is a mathematical technique for transforming information from the frequency domain into the time domain. The fast Fourier transform (FFT) is a computationally efficient technique for calculating the Fourier transform which exploits the symmetrical nature of the Fourier transform. If the Fourier transform is applied to the response, the following expression is obtained (Ewins 1995):

$$X(\omega) = \frac{1}{2\pi} \int_{-\infty}^{\infty} x(t)e^{-i\omega t} dt \qquad (A.1)$$

Similarly, the transformed excitation is (Ewins 1995):

$$F(\omega) = \frac{1}{2\pi} \int_{-\infty}^{\infty} f(t)e^{-i\omega t} dt \qquad (A.2)$$

The FRF $\alpha_{ij}(\omega)$ of the response at position i to the excitation at j is the ratio of the Fourier transform of the response to the transform of the excitation and is written as (Ewins 1995):

$$\alpha_{ij}(\omega) = \frac{X_i(\omega)}{F_j(\omega)} \qquad (A.3)$$

© Springer International Publishing Switzerland 2014
T. Marwala, *Artificial Intelligence Techniques for Rational Decision Making*,
Advanced Information and Knowledge Processing,
DOI 10.1007/978-3-319-11424-8

A.1.2   Wavelet Transform

The Wavelet Transform (WT) of a signal is an illustration of a timescale decomposition which highlights the local features of a signal. Wavelets occur in sets of functions that are defined by *dilation*, which controls the scaling parameter, and *translation*, which controls the position of a single function known as the *mother wavelet* $w(t)$. In general, each set of wavelets can be written as follows (Newland 1993; Marwala 2000, 2012):

$$W_{ab}(t) = \frac{1}{\sqrt{a}} w\left(\frac{t-b}{a}\right) \tag{A.4}$$

Here b=translation parameter, which localizes the wavelet function in the time domain; a=dilation parameter, defining the analyzing window stretching; and w=mother wavelet function. The continuous WT of a signal $x(t)$ is defined as (Newland 1993; Marwala 2000, 2012):

$$W\left(2^j + k\right) = 2^j \int_{-\infty}^{\infty} x(t)w^*(2^j t - k)dt \tag{A.5}$$

Here w^*=complex conjugate of the basic wavelet function; j is called the *level* (scale), which determines how many wavelets are needed to cover the mother wavelet and is the same as a frequency varying in harmonics and k determines the position of the wavelet and gives an indication of time. The length of the data in the time domain must be an integer power of two. The wavelets are organized into a sequence of levels 2^j, where j is from 1 to $n-1$. Equations A.4 and A.5 are valid for $0 \leq k$ and $0 \leq k \leq 2^j - 1$. The WT in this book is from the orthogonal wavelet family (Daubechie 1991) defined by Newland (1993) as follows (Marwala 2000, 2012):

$$w(t) = \frac{\left(e^{i4\pi t} - e^{i2\pi t}\right)}{i2\pi t} \tag{A.6}$$

The WT may also be formulated by transforming the signal $x(t)$ and the wavelet function into the frequency domain as follows (Marwala 2000):

$$W(j,k) = \int_{2\pi 2^j}^{4\pi 2^j} X(\omega)e^{i\omega k l 2^j} d\omega \tag{A.7}$$

The relationship between the physical properties of the structure and the WT of the impulse of unit magnitude may be applied to identify faults in structures. A functional mapping between the identity of a fault and the WT of the response k may be identified.

A.1.3 Modal Properties

This section reviews the modal properties which have been applied for damage identification in mechanical systems (Doebling et al. 1996). The modal properties are related to the physical properties of the structure. All elastic structures may be described in terms of their distributed mass, damping and stiffness matrices in the time domain through the following expression (Ewins 1995):

$$[M]\{\ddot{X}\}+[C]\{\dot{X}\}+[K]\{X\} = \{F\} \tag{A.8}$$

Here $\{X\}$, $\{\dot{X}\}$ and $\{\ddot{X}\}$ are the displacement, velocity and acceleration vectors respectively. $\{F\}$ is the applied force. If Eq. A.8 is transformed into the modal domain to form an eigenvalue equation for the ith mode, then (Ewins 1995):

$$\left(-\bar{\omega}^2[M]+ j\bar{\omega}_i[C]+[K]\right)\{\bar{\varphi}\}_i = \{0\} \tag{A.9}$$

Here $j = \sqrt{-1}$, $\bar{\omega}_i$ is the ith complex eigenvalue with its imaginary part corresponding to the natural frequency ω_i and is the ith complex mode shape vector with the real part corresponding to the normalized mode shape $\{\varphi\}_i$. The sensitivities of the modal properties for undamped case can be written to be (Ewins 1995):

$$\omega_{i,p} = \frac{1}{2\omega_i}\left[\{\varphi\}_i^T \left([K]_{,p} - \omega_i^2[M]_{,p}\right)\{\varphi\}_i\right] \tag{A.10}$$

and

$$\{\varphi\}_{i,p} = \sum_{r=1}^{N} \frac{\{\varphi\}_r\{\varphi\}_r^T}{\omega_i^2 - \omega_r^2}\left[[K]_{,p} - \omega_i^2[M]_{,p}\right]\{\varphi\}_i - \frac{1}{2}\{\varphi\}_i\{\varphi\}_i^T[M]_{,p}\{\varphi\}_i \tag{A.11}$$

In Eq. A.10 and A.11, N is the number of modes, $\omega_{i,p} = \dfrac{\partial\{\omega\}_i}{\partial g_p}$ $\varphi_{i,p} = \dfrac{\partial\{\varphi\}_i}{\partial g_p}$

$[K]_{mp} = \dfrac{\partial[K]}{\partial g_p}$, $[M]_{mp} = \dfrac{\partial[M]}{\partial g_p}$ and g_p represents changes in the pth structural parameters. The introduction of fault in structures changes the mass and stiffness matrices. Equations A.10 and A.11 show that changes in the mass and stiffness matrices cause changes in the modal properties of the structure.

A.1.4 Pseudo-Modal Energies

Another signals that is used in this book are pseudo-modal energies (Marwala 2001). Pseudo-modal energies are the integrals of the real and imaginary components of the frequency response functions over the chosen frequency ranges that bracket the natural frequencies. The frequency response functions may be expressed in receptance and inertance form (Ewins 1995). A *receptance* expression of the frequency response function is defined as the ratio of the Fourier transformed displacement to the Fourier transformed force; while the *inertance* expression of the frequency response function is defined as the ratio of the Fourier transformed acceleration to the Fourier transformed force. This section expresses the pseudo-modal energies in terms of receptance and inertance forms in the same way as the frequency response functions are expressed in these forms.

Receptance and Inertance Pseudo-Modal Energies

The frequency response functions may be expressed in terms of the modal properties by using the modal summation equation (Ewins 1995). Pseudo-modal energies may be estimated as a function of the modal properties from the frequency response functions expressed as a function of modal properties (Marwala 2001). This is performed in order to deduce the capabilities of pseudo-modal energies to identify faults from those of modal properties. The frequency response functions can be expressed in terms of the modal properties using the modal summation equation (Ewins 1995):

$$H_{kl}(\omega) = \sum_{i=1}^{N} \frac{\phi_k^i \phi_l^i}{-\omega^2 + 2j\varsigma_i \omega_i \omega + \omega_i^2} \qquad (A.12)$$

Here H_{kl} is the FRF due to excitation at k and measurement at l and ς_i is the damping ratio corresponding to the ith mode. Here it is assumed that the system is proportionally damped. This assumption is valid if the structure being analyzed is lightly damped. Proportional damping is defined as the situation where the viscous damping matrix $[C]$ is directly proportional to the stiffness $[K]$ or mass $[M]$ matrix or to the linear combination of both.

The Receptance pseudo Modal Energy (RME) is calculated by integrating the receptance FRF in Eq. A.12 as follows (Marwala 2001):

$$\begin{aligned} RME_{kl}^q &= \int_{a_q}^{b_q} H_{kl} d\omega \\ &= \int_{a_q}^{b_q} \sum_{i=1}^{N} \frac{\phi_k^i \phi_l^i}{-\omega^2 + 2j\varsigma_i \omega_i \omega + \omega_i^2} d\omega \end{aligned} \qquad (A.13)$$

In Eq. A.13, a_q and b_q represent respectively the lower and the upper frequency bounds for the qth pseudo-modal energy. The lower and upper frequency bounds bracket the qth natural frequency. By assuming a light damping ($\varsigma_i \ll 1$), Eq. A.13 is simplified to give (Gradshteyn and Yyzhik 1994; Marwala 2001, 2012)

$$RME_{kl}^q \approx \sum_{i=1}^{N} \frac{\phi_k^i \phi_l^i j}{\omega_i} \left\{ \arctan\left(\frac{-\varsigma_i \omega_i - jb_q}{\omega_i}\right) - \arctan\left(\frac{-\varsigma_i \omega_i - ja_q}{\omega_i}\right) \right\} \qquad (A.14)$$

B.1   Committee of Networks

B.1.1   Introduction

This section introduces the committee of networks which is used to build an ensemble. This technique was introduced by Perrone and Cooper (1992) and is illustrated in Fig. B.1.

The mapping of the input vector $\{x_k\}$ which represents the known input to the output vector y may be written as the desired function plus an error as follows (Perrone and Cooper 1992):

$$y_1 = h_1(\{x_k\}) + e_1(\{x_k\}) \tag{B.1}$$

$$y_2 = h_2(\{x_k\}) + e_2(\{x_k\}) \tag{B.2}$$

$$y_3 = h_3(\{x_k\}) + e_3(\{x_k\}) \tag{B.3}$$

In Eq. B.1—B.3 the approximated mapping function from network model is h_i and e_i is the mapping error vector for the ith network. The mean square error (MSE) for these three models in Eq. B.1, B.2 and B.3 may, therefore, be written as follows (Perrone and Cooper 1992):

$$E_1 = \varepsilon\left[(y_1 - h_1)^2\right] = \varepsilon\left[e_1^2\right] \tag{B.4}$$

$$E_1 = \varepsilon\left[(y_2 - h_2)^2\right] = \varepsilon\left[e_2^2\right] \tag{B.5}$$

$$E_1 = \varepsilon\left[(y_3 - h_3)^2\right] = \varepsilon\left[e_3^2\right] \tag{B.6}$$

In Eq. B.4 to B.6 the parameter ε indicates the expected value and corresponds to the integration over the input data and is defined as follows:

$$\varepsilon[e_1^2] = \int e_1^2 p(\{x_k\}) d\{x_k\} \tag{B.7}$$

$$\varepsilon[e_2^2] = \int e_2^2 p(\{x_k\}) d\{x_k\} \tag{B.8}$$

$$\varepsilon[e_3^2] = \int e_3^2 p(\{x_k\}) d\{x_k\} \tag{B.9}$$

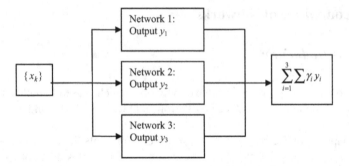

Fig. B.1 Illustration of a three network committee

In Eq. B.7—B.9, p is the probability density function and d is the differential operator. The average MSE of the three networks acting individually may thus be written as follows:

$$E_{AV} = \frac{E_1 + E_2 + E_3}{3} = \frac{1}{3}\left(\varepsilon\left(e_1^2\right) + \varepsilon\left(e_2^2\right) + \varepsilon\left(e_3^2\right)\right) \tag{B.10}$$

B.1.2 Equal Weights

In this section, the concept of a committee of networks is explained. The output of the committee is the average of the outputs from the three networks as illustrated in Fig. B.1. The committee prediction may thus be written in the following form by giving equal weighting functions:

$$y_{com} = \frac{1}{3}\left(y_1 + y_2 + y_3\right) \tag{B.11}$$

The mean square errors (MSE) of the committee of networks can be written as follows:

$$\begin{aligned}
E_{COM} \\
= \varepsilon\left[\left(\frac{1}{3}\{y_1 + y_2 + y_3\} - [h_1 + h_2 + h_3]\right)^2\right] \\
= \varepsilon\left[\left(\frac{1}{3}\left([(y_1 - h_1)] + [(y_2 - h_2)] + [(y_3 - h_3)]\right)\right)^2\right] \tag{B.12} \\
= \varepsilon\left[\left(\frac{1}{3}\left(e_1 + e_2 + e_3\right)\right)^2\right] \\
= \frac{1}{9}\left(\varepsilon\left[e_1^2\right] + 2\left(\varepsilon[e_1 e_2] + \varepsilon[e_1 e_3] + \varepsilon[e_2 e_3]\right) + \varepsilon\left[e_2^2\right] + \varepsilon\left[e_3^2\right]\right)
\end{aligned}$$

If it is assumed that the errors (e_1, e_2 and e_3) are uncorrelated then (Perrone and Cooper 1992):

$$\varepsilon[e_1e_2] = \varepsilon[e_1e_3] = \varepsilon[e_2e_3] = 0 \qquad (B.13)$$

To ensure that the errors are uncorrelated, in this chapter, a diverse committee of networks consisting of the multi-layer perceptron, radial basis functions and support vector machines is chosen. Substituting Eq. B.13 into Eq. B.12 then the MSE of the committee becomes (Perrone and Cooper 1992):

$$E_{COM} = \frac{1}{9}\left(\varepsilon\left[e_1^2\right] + \varepsilon\left[e_2^2\right] + \varepsilon\left[e_3^2\right]\right) \qquad (B.14)$$

The error of the committee in Eq. B.14 can be related to the average error of the networks acting individually (Eq. B.10) as follows:

$$E_{COM} = \frac{1}{3}E_{AV} \qquad (B.15)$$

Equation B.15 shows that the MSE of the committee is one-third of the average MSE of the individual method. From Eq. B.15, it can be deduced that the MSE of the committee is always equal to or less than the average MSE of the three methods acting individually.

B.1.3 *Variable Weights*

The three networks might not necessarily have the same predictive capacity. To accommodate the strength of each member of the committee, the network should be given appropriate weighting functions. It will be explained later how these weighting functions will be evaluated using the prior knowledge of the strength of each approach. The estimated missing data may be defined as the combination of the three independent approaches with approximate weighting functions as [a modification of Eq. B.11]:

$$y_{COM} = \gamma_1 y_1 + \gamma_2 y_2 + \gamma_3 y_3 \qquad (B.16)$$

In Eq. B.16, γ_1, γ_2 and γ_3 are the weighting functions and $\sum_{i=1}^{3} \gamma_i = 1$. The MSE due to the weighted committee can be written as follows (Perrone and Cooper 1992):

$$E_{COM}$$
$$= \varepsilon\left[\gamma_1\{x_u\}_1 + \gamma_2\{x_u\}_2 + \gamma_3\{x_u\}_3 - [\gamma_1 h_1 + \gamma_2 h_2 + \gamma_3 h_3]\right]$$
$$= \varepsilon\left[\left((\gamma_1\{x_u\}_1 - h_1) + (\gamma_2\{x_u\}_2 - h_2) + (\gamma_3\{x_u\}_3 - h_3)\right)^2\right] \qquad (B.17)$$
$$= \varepsilon\left[(\gamma_1 e_1 + \gamma_2 e_2 + \gamma_3 e_3)^2\right]$$

Equation B.17 may be rewritten in Lagrangian form as:

$$E_{COM} = \varepsilon[\gamma_1 e_1 + \gamma_2 e_2 + \gamma_3 e_3] + \lambda(1 - \gamma_1 - \gamma_2 - \gamma_3) \qquad (B.18)$$

Here λ is the Lagrangian multiplier. The derivative of error in Eq. B.18 with respect to γ_1, γ_2, γ_3 and λ may be calculated and equated to zero as follows:

$$\frac{dE_{COM}}{d\gamma_1} = \varepsilon\left[2(\gamma_1[e_1] + \gamma_2[e_2] + \gamma_3[e_3])[e_1]\right] - \lambda = 0 \qquad (B.19)$$

$$\frac{dE_{COM}}{d\gamma_2} = \varepsilon\left[2(\gamma_1[e_1] + \gamma_2[e_2] + \gamma_3[e_3])[e_2]\right] - \lambda = 0 \qquad (B.20)$$

$$\frac{dE_{COM}}{d\gamma_2} = \varepsilon\left[2(\gamma_1[e_1] + \gamma_2[e_2] + \gamma_3[e_3])[e_3]\right] - \lambda = 0 \qquad (B.21)$$

$$\frac{dE_{COM}}{d\lambda} = (1 - \gamma_1 - \gamma_2 - \gamma_3) = 0 \qquad (B.22)$$

Solving Eq. B.19—B.22, the minimum errors obtained are:

$$\gamma_1 = \cfrac{1}{1 + \cfrac{\varepsilon[e_1^2]}{\varepsilon[e_2^2]} + \cfrac{\varepsilon[e_1^2]}{\varepsilon[e_3^2]}} \qquad (B.23)$$

$$\gamma_2 = \cfrac{1}{1 + \cfrac{\varepsilon[e_2^2]}{\varepsilon[e_1^2]} + \cfrac{\varepsilon[e_2^2]}{\varepsilon[e_3^2]}} \qquad (B.24)$$

$$\gamma_3 = \frac{1}{1 + \frac{\varepsilon[e_3^2]}{\varepsilon[e_1^2]} + \frac{\varepsilon[e_3^2]}{\varepsilon[e_2^2]}} \qquad \text{(B.25)}$$

Equations B.23—B.25 may be generalized for a committee with n-trained networks and be expressed for network i as follows:

$$\gamma_i = \frac{1}{\sum_{j=1}^{n} \frac{\varepsilon[e_i^2]}{\varepsilon[e_j^2]}} \qquad \text{(B.26)}$$

By analyzing Eq. B.26, it is deduced that if the predictive capacity of the three networks are equal, then each method should be assigned equal weights. This conclusion is trivial, but it is deduced in this chapter to confirm the effectiveness of the proposed method. Since it is not known which network is more accurate at a given instance, the weighting functions are determined from the data that is used for training and validation of the networks process (prior knowledge).

B.1.4 Committee Gives More Reliable Solution

Axiom: If three independent (uncorrelated) methods are used simultaneously, the reliability of the combination is at least as good as when the methods are used individually. Suppose the probabilities of success for the network 1 (m_1), network 2 (m_2), and network 3 (m_3) are $P(m_1)$, $P(m_2)$, and $P(m_3)$, respectively. The reliability of the three methods acting in parallel is given by (Marwala 2001):

$$
\begin{aligned}
P(m_1 \cup m_2 \cup ...m_n) = {} & P(m_1) + P(m_2)... + P(m_n) \\
& - \left[\begin{matrix} P(m_1 \cap m_2) + P(m_2 \cap m_3) + ... \\ + P(m_{n-1} \cap m_n) \end{matrix} \right]... \\
& + P(m_1 \cap m_2...m_n)
\end{aligned}
\qquad \text{(B.27)}
$$

From Eq. B.27, it can be deduced that the reliability of the committee is always higher than that of the individual methods.

References

Doebling SW, Farrar CR, Prime MB, Shevitz DW (1996) Damage identification and health monitoring of structural and mechanical systems from changes in their vibration characteristics: a literature review Los Alamos National Laboratory Report LA-13070-MS

Ewins DJ (1995) Modal testing: theory and practice. Research Studies Press, Letchworth

Gradshteyn IS, Yyzhik IM (1994) Tables of integrals, series, and products. Academic Press, London

Marwala T (2000) On damage identification using a committee of neural networks. American Society of Civil Engineers, J Eng Mech 126:43–50

Marwala T (2001) Fault identification using neural networks and vibration data. Doctor of Philosophy, University of Cambridge

Newland DE (1993) An introduction to random vibration, spectral and wavelet analysis, 3rd edn. Longman, New York

Perrone MP, Cooper LN (1992) When networks disagree: ensemble methods for hybrid neural networks. Brown University, Technical Report ADA260045

Index

© Springer International Publishing Switzerland 2014
T. Marwala, *Artificial Intelligence Techniques for Rational Decision Making,*
Advanced Information and Knowledge Processing,
DOI 10.1007/978-3-319-11424-8